true green @ work

true green @ work

100 ways you can make the environment your business

Kim McKay and Jenny Bonnin
with Tim Wallace

NATIONAL GEOGRAPHIC

WASHINGTON, D.C.

Published by the National Geographic Society

First published in Australia by ABC Books in August 2007 for the
Australian Broadcasting Corporation.
Copyright © 2008 True Green (Global) Pty Ltd .

ISBN: 978-1-4262-0236-6

Library of Congress Cataloging-in-Publication Data available upon request.

Design, layout, and select images by Marian Kyte
Written by Tim Wallace

A percentage of proceeds from the sale of *True Green @ Work*
benefits Clean Up the World.

True Green @ Work has purchased carbon credits to neutralize
emissions produced by the printing of this book.

True Green ® is a Trademark of True Green (Global) Pty Ltd.

Founded in 1888, the National Geographic Society is one of the largest nonprofit
scientific and educational organizations in the world. It reaches more than 285 million
people worldwide each month through its official journal, NATIONAL GEOGRAPHIC, and its
four other magazines; the National Geographic Channel; television documentaries;
radio programs; films; books; videos and DVDs; maps; and interactive media. National
Geographic has funded more than 8,000 scientific research projects and supports an
education program combating geographic illiteracy.

For more information, please call
1-800-NGS LINE (647-5463)
or write to the following address:

National Geographic Society
1145 17th Street N.W.
Washington, D.C. 20036-4688 U.S.A.

Visit us online at
www.nationalgeographic.com/books

Printed in Italy on recycled paper.

contents

foreword   7

introduction   8

desk   11   ●

office   23   ●

building   35   ●

culture   47   ●

stakeholders   59   ●

carbon neutral 71   ●

closing the loop 83 ●

ecolabeling   95   ●

marketing   107   ●

green business 119   ●

resources   **131**

glossary   **136**

sustainability @ work

22   general electric

34   visy

46   sc johnson

58   marriott

70   office depot

82   google

94   veolia

106 timberland

118 patagonia

130 interface

foreword

Wendy Gordon
Founder, The Green Guide

Imagine the impact on the book industry when the publisher of Harry Potter decided to print every copy of J.K. Rowling's last book on Forest Stewardship Council-Certified paper made from sustainably managed forests. Or think what it means when Fed Ex and a growing number of cities choose to convert portions of their truck fleets to hybrid engines, or Wal-Mart requires manufacturers to reduce their products' packaging or risk losing their contracts.

Of course, most of us aren't in positions where we can make those sorts of decisions. But no matter what our role in the workforce, or the size of our workplace, there are countless ways we can make the green difference. How? Going green saving energy and water, which saves money. But more importantly, going green at work means less fossil fuel burning, less greenhouse gases released, and less global warming, fewer chemicals in the waste stream, more trees and cleaner air. Call it the planet's bottom line.

I got a call not too long ago from a friend who works in a giant multimedia company. He was asking if the Green Guide might help them green their offices. I said, "Sure, I've got just what you need: this great new book—*True Green @ Work*. Everything you're looking for to green your workplace is right here in this compact guide."

It's not always easy when you've got deadlines and distractions to be thinking about greener ways to work. That's why we all need *True Green @ Work*. There are so many simple things we could be doing to save resources that probably never crossed our minds—from setting the printer to make double-sided copies, to recycling ink cartridges and teleconferencing instead of expending money and emitting CO_2 to travel to a meeting.

Our small office, a satellite office of the National Geographic, is making changes in the paper we buy. While reducing paper use is priority number one, what paper we generate we want to be sure is recycled. This has meant discussions with the building management manager and the city's environmental agency.

Now is the time. Cities around the U.S. and businesses are engaged in some very exciting new thinking. Cities are competing to be the greenest and offering incentives to businesses that want to be greener. Businesses meanwhile are innovating at an unprecedented pace to change how their products are made, packaged, stored, shipped, used, and disposed. They're looking at every aspect of their operation from the manufacturing plant to the supply room, from the cafeteria to the executive suites. Doing good and being good, most executives will tell you today, work hand in hand.

Americans are hardworking people, most putting in more than a third of each day at work, five days a week for dozens of years. Not all workplaces are conducive to high productivity, however, or good health. It may surprise some that serious indoor air pollution, "sick building syndrome," affects many offices, thanks to poor ventilation and toxins off-gassing from the rugs, cabinets, and electronic equipment. A greener office is a cleaner, safer office designed with people in mind—one that's energy and water efficient, free of indoor air pollutants, flooded with natural light, a place that's easy to get to by public transportation or even on foot or bicycle. *True Green @ Work* can help you make your workplace a greener, healthier place to spend your days.

So take *True Green @ Work* with you when you head to work. Talk with your colleagues, put together a committee, conduct an audit of your workplace to determine the issues and the solutions. This book can be your guide. Its 100 steps are practical and manageable to implement; its creators, Kim McKay and Jenny Bonnin, a True Green inspiration.

introduction

Kim McKay and Jenny Bonnin

We spend about a third of our lives at work, with Americans clocking some of the longest hours in the developed world.

Our working environment has been shown to contribute significantly to how we feel about ourselves, how we conduct relationships with our peers and families and whether we lead a healthy and fulfilling life. While work is just one aspect of life, the satisfaction derived from a job well done can impact on your whole well-being; it can also be financially rewarding.

However, the extra time we are putting in might not be in our own best interests.

Workplaces, after all, contribute to the consumption of large quantities of energy, water, and other resources, producing greenhouse gas emissions in the process; yet nearly seven out of 10 of us believe action to combat climate change is critical to the world's vital interests.

Given all this time spent at work, combined with our increasing concern about the effects of global warming, it's no wonder that people want to know what they can do to help make their workplace more environmentally friendly and their companies more sustainable.

Since releasing *True Green: 100 Everyday Ways You Can Contribute to a Healthier Planet* (National Geographic Books, 2007), we have been inundated with questions from readers about what more they can do at work to make a difference.

In researching the concept for this book and looking for practical examples of green businesses, we found many companies that have implemented significant sustainability practices over the past few years; but the number is still only a comparative handful.

Sometimes it can seem hard to make a difference when you aren't in charge; sometimes it seems hard even when you are. But from our own experience with Clean Up the World we have seen how everyone has an important role to play in making things happen.

From the chief executive implementing across-the-board sustainability practices and the factory-floor supervisor who comes up with an innovative way to recycle offcuts to the junior clerk who oversees recycling in the staff lunch room— it all makes a difference to the final outcome.

Plus it makes good business sense.

Consider the weight of evidence of why business needs to act now. The iconic global insurance company Lloyd's of London has named climate change one of the biggest threats facing global business alongside terrorism, political instability, and natural catastrophes.

< our challenge is to every business: step up and take your environmental responsibility seriously >

Lloyd's chairman Lord Levene put it this way: "Failure to take climate change into account will put companies at risk from future legal actions from their own shareholders, their investors and clients. Climate change must inform underwriting strategy—from the pricing of risk to the wording of policies. It must guide and counsel business strategy—including business development and planning."

Economist Sir Nicholas Stern concluded in his landmark 2006 report for the British Government that 1 percent of global gross domestic product needs to be invested to mitigate the effects of climate change, and that failure to do so risks global GDP being up to 20 percent lower than it otherwise might be. Without action, his report says, the costs will include dealing with up to 200 million environmental refugees, displaced by both desertification on the one hand and rising seas on the other.

The man who commissioned Stern's report, the now British prime minister, Gordon Brown, has called the green challenge an opportunity for "new markets, new jobs, new technologies [and] new exports where companies, universities and social enterprises [in Britain] can lead the world."

Corporate social responsibility isn't just another management jargon term or the marketing trend of the month that you can pay lip service to and ignore.

If you want your company to be around beyond the next quarterly results it is the only way forward. It is about protecting and preserving what we have for future generations and their right to work and live in harmony with the environment.

We're pleased to be associated with National Geographic, which has been inspiring people to care about the planet for 120 years. It too has embarked on a program to implement sustainable practices across the organization and reduce carbon emissions while continuing to inform the global community about the critical environmental issues we all face.

We hope you'll help put into practice as many of the 100 ideas in *True Green @ Work* as you can, learning from the experiences of the 10 sustainability champions featured in the book. It's not difficult; all it takes is a little effort and a commitment to live your life in a true green way.

Go to our website, www.betruegreen.com, and tell us what you are doing at work to make a difference.

Our challenge is to every business leader to step up and take your environmental responsibility seriously. Your shareholders care, your suppliers and customer care, and your workforce most certainly cares.

desk

1   coffee fix

Small things can make a big difference. Consider your humble cup of coffee, the world's second-most valuable legal commodity after oil, with environmental and social impacts to match. Stir up some positive change by asking your barista and/or office manager to switch to organic and Fair Trade-labeled brands. Then do your bit to reduce the energy and waste involved in producing, transporting, and disposing of more than 1.9 million tons of paper and plastic cups and plates thrown away in the U.S. each year: use your own mug. You need to wash it, sure, but life cycle analysis shows that over its life of about 3,000 uses a mug is associated with 30 times less solid waste and 60 times less air pollution than the equivalent drinks in disposable paper or foam cups.

pens and pencils

2

Americans buy more than 5.1 billion pens every year, and most are the disposable type, thrown in the trash once the ink runs dry. That adds up to about 770 tons more of unnecessary plastic waste in landfills each year. Make a better impression in your written work by using long-life refillable pens made from recycled plastic, paper, or timber, or from fully biodegradable bioplastic (derived from corn starch). Also look for pencils manufactured from sustainably harvested timber or wood substitutes such as recycled paper, old plastic cups, wood offcuts, and reclaimed denim.

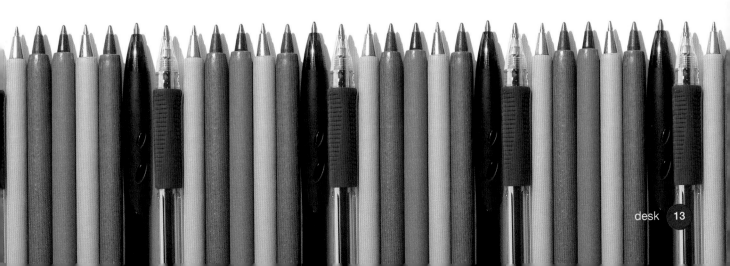

3

paper

Paper is up to 70 percent of a typical office's waste, so simple measures to reduce unnecessary use can significantly lower your organization's waste removal expenses—not to mention purchasing costs. Use both sides. Keep a tray on your desk to collect any single-side printed scrap paper that comes your way and use it for note paper, or in the photocopier or fax machine. Format documents to avoid printing unnecessary pages and proofread carefully on screen to avoid printing multiple copies. Keep a paper recycling box next to your desk and encourage colleagues to separate paper from organic waste and recyclable containers.

other stationery

4

What's true of pens and paper is true of other office consumables, with a wide variety of envelopes, sticky notes, ring binders, notebooks, and other general supplies made from recycled and eco-friendlier materials now available. Use a desk caddy to save envelopes, rubber bands, and paper clips for reuse, and ask your office manager to establish an exchange spot for other items such as bubble wrap and padded envelopes. Request designs that support reuse, such as envelopes that have resealable flaps and space for multiple addresses, or adhesive-tape holders that are refillable rather than disposable.

Supporting the environment – 100% Recycled Paper

5

sleep more

Between meetings, coffee breaks, phone calls, and lunch, the chances are that in any given working day there's at least an hour where you don't actually use your computer. By activating the sleep mode you can reduce its electricity consumption to less than 5 percent of full power. Go into system preferences and set them to put the screen and hard drive to sleep after 10 minutes of inactivity, ensuring savings when you're unexpectedly caught up doing something else. Detailed guides on how to enable energy-saving features on Windows, OS/2, Unix, and Mac OS X systems are available from the U.S. Environmental Protection Agency's Energy Star website: www.energystar.gov.

6

correspondence

From direct-mail catalogs and conference invitations to bulk faxes and form letters, the amount of unwanted printed correspondence you receive often outweighs that which is relevant—wasting not only resources but also your time in having to sort through it. Resist the urge just to throw it straight into the recycling bin; that won't solve the problem in the long term. Make the effort to remove you or your business from mailing lists by contacting the Direct Marketing Association or individual companies; if that's not an option, write "not at this address, return to sender" and put it back in the mail.

7

food for thought

Buying your lunch is arguably more water- and energy-efficient than making it yourself—particularly if it involves heating—but a home-packed lunch is undoubtedly cheaper and produces less solid waste than takeout food. It also gives you greater opportunity to eat organic produce, which is better for the environment, contains fewer chemical traces, and has higher nutritional value—plus it tastes better. Rather than buying plastic containers, reuse takeout containers before throwing them into the recycling bin. Reuse bread wrappers and other plastic packaging rather than buying plastic wrap or aluminum foil. Keep reusable spoons, forks, and chopsticks in your drawer so you don't need the disposable kind.

plant life

Indoor plants can play a crucial role in
your local work environment. A plant on
your desk is not only nice to look at but
also acts as a natural air filter, absorbing
airborne pollutants and computer radiation
while replenishing oxygen levels. It's also an
air cooler, through the evaporative process
known as transpiration. Indoor plants
help protect you from the germs of your
colleagues, with research showing they
significantly reduce the incidence of fatigue,
coughs, sore throats, and other cold-related
illnesses. Plus they have a measurable effect
in reducing stress levels, so it doesn't hurt
to have one close by for when you get stuck
on hold while on the phone.

9 dry-cleaning

You think you're picking up fresh, clean work clothes, but you might also be getting a sniff of dirty environmental laundry. Dry cleaners use large amounts of the chemical solvent tetrachloroethylene, a powerful degreasing agent that's also a suspected carcinogen, can aggravate asthma and allergies, and is harmful to the environment. During its production, transport, and use, tetrachloroethylene breaks down into chemicals such as the toxin phosgene and contributes to photochemical smog. Before opting for dry cleaning, consider a quick cold hand wash or spot cleaning. Look for a service with "clean and green" processes, including reuse of hangers and garment bags.

shut down

The idea that leaving a machine on is more efficient than turning it off has become something of an urban, and deeply uneconomic, myth. Left on all day, every day, as happens in some offices, a computer will over a year use nearly 1,000 kilowatts of electricity, resulting in more than a ton of carbon emissions and an unnecessarily high electricity bill. By switching off your computer before you go home you'll cut its electricity use to less than 250 kilowatts, with comparable carbon and cost savings. Think about turning it off even when you're going to a meeting or lunch. Do the same with other office equipment.

< GE understands that through the use of market-based policies, we can spark the transition to a low-carbon economy that will be good for the company, good for the economy, and good for the planet >

GENERAL ELECTRIC, **born out of Thomas Edison's electric light company over 125 years ago, has turned its energies to creating a greener future. With one of the largest arrays of infrastructure and industrial products, services, and financing in the world, GE has committed to innovative solutions for environmental challenges through "Ecomagination,"—a business strategy to meet customers' demands for more energy-efficient, cleaner products.**

inspiration> Through Ecomagination, GE has pledged to decrease pollution from its products and to double research and development spending on cleaner technologies. Using a series of "dreaming sessions," first held in 2005, the company has engaged its customers in helping to shape the company's visionary strategy.

achievements> GE has developed an Ecomagination Product Review scorecard, quantifying a product's environmental impacts and benefits relative to other products on the market. To ensure the scorecard's accuracy, GE sought outside help to provide independent, quantitative environmental analysis and verification of its product claims.

After recommending lighting retrofits to customers as a way to cut energy spending, GE took its own advice, rolling out a two-year plan to retrofit lighting at 148 of its industrial and manufacturing warehouses worldwide—an effort that could cut annual lighting energy costs at each facility by an average of 50 percent.

staff> GE has launched a company-wide communication campaign to engage its 300,000 employees and locations in greenhouse gas reduction efforts. In doing so, the company hopes to ensure that clean technology and environmental stewardship are part of everyone's job over the long term.

partnerships> The company hopes to increase its engagement with the public through the Ecomagination Advisory Council, a board of six to eight industry thought leaders with expertise in energy and the environment.

initiatives> GE is investing in tomorrow's energy technologies—from renewable energy to hydrogen, and from lightbulbs to aircraft engines and household appliances. Among the company's technology initiatives are sophisticated wind turbines and solar panels, more effective use of biofuels in its engines, and distributed energy generation systems to provide more cost effective solutions using hybrid power generation.

challenges> GE has set a goal to reduce its absolute greenhouse gas emissions by 1 percent worldwide by 2012—admittedly a big goal for a company whose emissions would otherwise have grown substantially based on current business growth projections.

what's ahead> Looking 10 years ahead and beyond, GE researchers are working to commercialize organic light-emitting diode lighting applications that will provide customers with an entirely different way to light homes and businesses. Such applications, along with its growing line of green initiatives, position GE to be a global green technology leader for years to come.

Sourced from www.ge.com

office

team effort

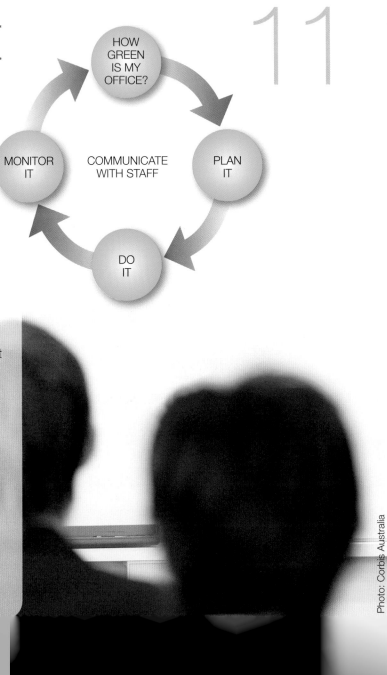

The cycle diagram shows: HOW GREEN IS MY OFFICE? → PLAN IT → DO IT → MONITOR IT → (back to HOW GREEN IS MY OFFICE?), with COMMUNICATE WITH STAFF at the center.

A green business begins with personal commitment. Without individuals doing the right thing—whether at their desk, on the shop floor, or out on the road—no environmental policy will work. But the inverse is also true: what any individual can do is limited without a team approach. To achieve optimum environmental and financial outcomes you need shared goodwill plus workable systems. Help create a green office culture with a management plan that harnesses the commitment of everyone in your workplace—starting with simple things like making recycling easy by ensuring everyone has collection bins at their desks or work stations. Provide briefings, signs, and other information to explain and encourage ways to conserve energy and minimize waste.

Photo: Corbis Australia

12

measure and manage

Commercial buildings and industrial facilities produce 45 percent of the nation's greenhouse emissions. Energy use in intensive factory operations is understandable; in office buildings it's unnecessarily high. Lease arrangements often provide no financial incentive for a tenant to introduce energy-efficiency measures, nor even the capability to monitor use. Negotiate a lease in which reduced energy and water consumption will benefit both your company and your landlord. Managing building-related electricity use can be particularly cost effective since it makes up a disproportionately large part of peak electricity demand, which can amount to as much as half of total electricity costs.

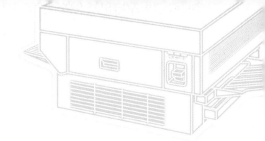

13

office machines

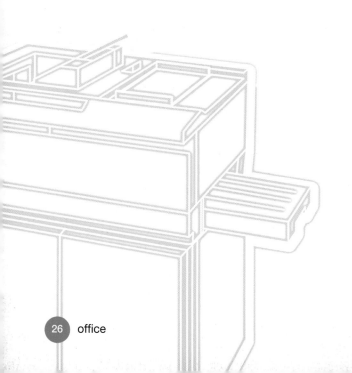

Energy-efficient office machines not only cut the cost and carbon emissions associated with their own operation but, through producing less heat, they can reduce emissions from air conditioning by as much as 30 percent. Most energy in photocopiers and printers is used to heat the components that fuse the toner to paper, so a machine that powers down rather than keeping those components hot while idle can slash energy use. Ensure your procurement policy favors Energy Star-compliant and enabled equipment. Choose models that allow double-sided printing and copying, since ultimately the greatest source of carbon emissions from office machines is the embedded energy of consumables like paper and ink.

14
computers

Computers are, on a weight for weight basis, more environmentally damaging to produce than cars. Research by the UN University in Tokyo indicates manufacturing of a standard desktop computer and 17-inch CRT monitor requires a total of two tons of materials, including at least 528 pounds of fossil fuels, 48 pounds of chemicals, and nearly 400 gallons of water. A good place to start when choosing machines is the Green Electronic Council's Electronic Product Environmental Assessment Tool, which rates computer companies on their material use and end-of-life take-back policies. Minimize the need to replace computers so often by choosing models whose life can be extended through upgrades and repair.

Photo: Corbis Australia

office paper 15

Despite advances in technology, the paperless office remains a futuristic fantasy, with about 10,000 sheets of letter paper—as much paper as is produced from pulping a full-grown tree—being used for every U.S. worker each year. Much of this paper comes from native forests and is chlorine-bleached, a process that produces toxic dioxins. Recycled paper uses up to 90 percent less water and half the energy required to make paper from virgin timber, yet less than 10 percent of the over 12 million tons of printing and writing paper used in the U.S. each year is recycled content. Do the paperwork: you'll often find it hard to spot the difference between recycled and virgin paper, with manufacturers guaranteeing recycled paper for virtually all office functions.

cartridges 16

Dire warnings against reusing printer ink and toner cartridges help protect the profit margins of manufacturers but contribute to more than 167 million environmentally unfriendly cartridges being dumped in U.S. landfills each year, generating 40 million pounds of unnecessary waste. Subject to the fine print on a warranty, however, there's no good reason a cartridge can't be reused up to four times. It will not only cut waste but also save up to 90 percent on the cost of a new cartridge. Use a refiller or remanufacturer prepared to offer a written guarantee against equipment damage and downtime caused by their products. When replacing a printer or copier, consider buying one with a long-lasting print drum that requires only toner refills.

furnishings

Traditional office furnishings not only are resource-intensive but can increase office air pollution through the emission of volatile organic compounds from their glues and finishes. Clean up your immediate environment and reduce your global impact by choosing eco-friendly furniture and floor coverings. These are made from recycled or sustainably harvested nontoxic materials and designed so they can be dissembled for reuse at the end of their life. You can find chairs, for example, made from recovered railway cars, old phones, computer casings, and even vacuum cleaners. Or save the resources required to make new items by buying refurbished furniture.

Mitt armchairs made with Recopol™ recycled shells.
Photo: Wharington International, manufacturer

kitchen facilities

18

Good kitchen facilities allow staff to bring and prepare their own food from home, which is usually cheaper and healthier, and produces less solid waste than buying takeout food. Your office kitchen should have separate bins for recycling plastic, glass, and aluminum containers, along with one for organic matter such as food waste, which can be diverted from landfills by being converted on-site into compost. Avoid the long-term cost of disposable paper, plastic, or foam cups, plates, and utensils by investing in the real thing—or choose eco-friendly alternatives made from recycled plastic, biodegradable bamboo, or bioplastics such as corn or potato starch. If your office provides coffee and tea or cocoa, choose organic and Fair Trade brands.

Photo: Corbis Australia

tissue paper

19

Napkins, toilet paper, and paper bathroom and kitchen towels can make up as much as a third of the volume of office waste. Not only are many of these products made from virgin timber, but like office paper they are bleached in a process that emits dioxins. Choose products made from recycled paper or tree-free alternatives such as kenaf or hemp fiber. With advances in technology, recycled products are just as strong, soft, absorbent, hygienic, and even aesthetically pleasing. Reusable and laundered roll towels can be cheaper than paper towels, and make for less waste in your bathroom. Hand dryers are another possible option, although they add to your energy bill and have a debatable hygiene advantage.

cleaning agents

The very products and processes used to keep indoor environments clean may also contribute to indoor pollution. Studies show that in certain conditions many common products used for cleaning floors, surfaces, and plumbing fixtures contain toxic contaminants at levels that pose risks to both human health and the environment. Symptoms can include eye, nose, and throat irritation along with headaches, dizziness, and fatigue. Long-term environmental consequences, such as contamination of surface and ground water and bioaccumulation in plants and animals, can also occur when these products are poured down drains. Use a cleaning contractor committed to clean and green products and processes.

< we have grown by providing services for the reuse and recycling of our products >

VISY began as a box-making business and has grown to become the world's largest privately owned packaging and recycling company, with annual revenue in excess of $2.6 billion and more than 5,000 staff. It now collects more than 1.6 million tons of recyclable materials. Tony Gray is group public affairs manager.

inspiration> Our investment in recycling has given us the opportunity to differentiate our business and gain an advantage in our market, where issues such as product stewardship and landfill avoidance are being discussed throughout the supply chain. Plus it is the right thing to do.

achievements> Development of a dedicated "Sustainability Services" business unit to implement reuse and recycling solutions for our products once they reach the end of their life. More than 70 staff now work in the unit in Australia, and the numbers are expanding, providing more solutions to our customers.

management> To change habits and company ways management adopted our noble purpose—"to help people make sustainable choices"—along with a set of values and goals for all employees. After seeing the results of our sustainability practices and becoming more aware, our people have embraced sustainability goals.

staff> Sustainability underpins all of our practices and is endorsed throughout the business. Surveys show staff support our sustainability goals.

suppliers> Our procurement division works with our suppliers to provide us with more efficient products and services aligned with our sustainability goals. This has resulted in better economic outcomes.

government> Governments have been slow to support businesses on their sustainability routes but are gradually seeing the necessity to do so and starting to change policies.

costs> We have grown our business by providing new and improved services.

benefits> Our traditional manufacturing sites use resources more efficiently and generate less waste, meaning direct savings to our bottom line. We are always learning.

initiatives> Asking our staff for their feedback on our sustainability goals via surveys has provided us with a lot of great suggestions, as well as challenges we need to overcome as we move our business further down the sustainability path.

challenges> There was no resistance from the board or management, though there was skepticism and a lack of understanding about the importance of the business unit and opportunities it would create.

advice> 1> Establish a long-term vision and goals that make sense for your business, taking into account your sphere of influence. 2> Ensure the vision and goals are considered in every part of your business operations. 3> Communicate your goals and progress to employees, customers, and other stakeholders.

building

star quality

A green building is a high-performance building. Inspired by the efficiency of nature, it makes the most of passive (low-energy) technology to create a pleasant interior environment while reducing its impact on the wider environment. As well as reaping lower costs from energy efficiency, water conservation, and waste minimization, companies also cite significant productivity benefits from healthier, happier staff. Occupant well being is just as important to rating a green building as calculating its carbon emissions. For more information on the U.S. Green Building Council's Leadership in Energy and Environmental Design, or LEED rating system for commercial office design, construction, and refurbishment, visit www.usgbc.org.

green from go

Designing a green workplace begins with effective management, taking a whole-of-building approach that accounts for the interaction between structure, systems, and external environment. Simple measures, when integrated in a comprehensive plan, can contribute to significant energy savings. A sensor system to control lighting, for instance, also reduces heat generation and the need for air conditioning. The potential of this holistic approach is shown in buildings such as the City of Chicago's Center for Green Technology, one of the nation's first buildings certified platinum, the highest achievable level, by the U.S. Green Building Council. Along with comprehensive greenhouse and waste management plans, green features of the building include rooftop solar energy panels, rainwater collection for irrigation, recycled building materials, sensors to control lighting, and a vegetated roof.

location, location, location

23

Buildings produce carbon emissions during construction, refurbishment, and eventual demolition, from embodied energy in building materials and furnishings to the operational demand of lighting, ventilation, air conditioning, and other systems. But even before workers walk through the door, they contribute emissions through travel—estimated at an average of 2.2 tons of carbon dioxide from each worker every year. Plan your building to encourage public and alternative transportation. The Lake View Terrace Library in Los Angeles, certified platinum by the U.S. Green Building Council, provides safe access for bicyclists and pedestrians, a bicycle storage area, easy access to public transportation—even a place to charge electric vehicles.

breathing space

24

Ventilation accounts for more than 20 percent of commercial building greenhouse gas emissions. That's because buildings reliant on artificial air conditioning achieve "energy efficiency" by sealing off the internal from the outside environment and then using machines to push the air around. Studies show indoor air quality can be many times worse than outdoors, leading to a range of health problems—including dizziness, nausea, and fatigue—known as sick building syndrome. By using natural ventilation to regulate the temperature, a green building is a healthier building. A good example is Sidwell Friends Middle School in Washington, D.C., where operable windows, skylights, and ceiling fans maintain comfort for students and teachers while minimizing the need for artificial cooling.

25

going with the flow

Green buildings achieve a comfortable interior environment without energy-intensive air conditioning and heating (which together account for more than 40 percent of commercial building greenhouse emissions). Some of the technologies are as old as architecture, from site orientation and window placement to using reflective barriers and thermal mass. Others are more recent, such as passive chilled beams—water-cooled tubes or panels that exploit natural physics to absorb heat and then dissipate it through cooling towers or thermal chimneys. Chilled beams were first used in the 1960s but took a few decades to catch on; now they're becoming commonplace, thanks to projects such as the Oregon Health and Science University's Center for Health and Healing in Portland, the first large building in the U.S. to use radiant chilled beams to supplement air conditioning.

26

landscape views

Using the landscape and planting vegetation to create habitat for other flora and fauna improves a green building's appearance and provides tangible benefits in environmental management. Statesville, North Carolina's Third Creek Elementary School uses a natural wetland to slow and filter storm water runoff before it reaches a nearby stream, also providing students with an outdoor classroom and living wildlife laboratory. From New York high rise buildings to the rooftops of Seattle libraries, vegetated, or "green," roofs are also sprouting up, serving as a thermal insulating mass and helping to manage rainwater runoff. Installing a rooftop garden costs more, but a green roof will also last years longer than a standard roof.

generation now

27

A green building seeks to meet its own energy needs. It can be as simple as using solar energy to heat water, though producing electricity directly from photovoltaic cells is also an increasingly viable option, both in the U.S. and overseas. Integrated wind turbines and photovoltaic grids helped the CH2 project in Melbourne, Australia, achieve an impressive six-star rating from the Green Building Council of Australia. It's anticipated that greenhouse emissions produced by the 10-story building, to be occupied by City of Melbourne offices, will be 87 percent lower than the building it replaces and require just one-seventh of the electricity and gas. Out of a total projected budget of $34 million, $9.5 million is being invested in sustainability features, but that's expected to pay for itself in 10 years.

CH2 Melbourne

28 liquid assets

With water supplies becoming an ever-increasing concern, businesses will likely face paying considerably more for water in the future. Cost-effective measures include water-conserving fixtures and plumbing such as reduced-flow taps, waterless urinals (which are more sanitary than standard ones), and low-flush toilets; using native and drought-tolerant plants in landscaping; irrigating with water recycled from sinks, basins, and toilets; collecting rainwater for on-site use, and using green roofs, ponds, and other water-catching features to manage rainwater runoff. The Sweetwater Creek State Park Visitor Center in Lithia Springs, Georgia, incorporates composting toilets, waterless urinals, and a rainwater capture and reuse plan into its design, reducing water use by 77 percent.

29

material gains

While greenhouse gas emissions from the embodied energy used in building materials are about an eighth of those from the operational energy a commercial building uses over a life span of 40 years, they are rising. This is due to the use of more energy-intensive materials (stainless steel and highly coated glass, for instance) in construction; the increasing size of buildings; more frequent refurbishment; and ever more machine-intensive building methods. A green building seeks to use recycled materials and to recycle materials. The University of California, Santa Barbara's Bren Hall recycled 100 percent of the demolition waste from the parking lot it replaced onsite or elsewhere on campus—even down to the former parking lot's landscaping and soil. Builders also recycled over 93 percent of construction waste.

in sight, in mind

The best-designed green building is only as good as its occupants. Portland, Oregon's Jean Vollum Natural Capital Center houses a mix of nonprofit and business tenants gathered around the themes of sustainable forestry and fisheries, green building, and financial investment. The building invites and engages public curiosity through generous public spaces for exhibits, events, and spontaneous interactions. Interpretive "niches"—display spaces made of wood, cork, and piping salvaged from the original building—are recessed into walls, providing education displays and encouraging supportive behavior. Displays are rotated every two to three months and offer visitors information on green building, rainwater management, sustainable forestry, and several other topics.

courtesy Ecotrust

< winning is not just about performance and success; it is also about doing what's right—right for the planet, right for people, right for the next generation >

SC JOHNSON **was founded in 1886 by Samuel Curtis Johnson as a parquet flooring company. It's become a global household name and one of the world's leading manufacturers of cleaning products and products for home storage, air care, personal care, and pest control. Now, the company is not only striving to make consumers' lives cleaner, easier, and healthier, but to build a cleaner, greener future by improving the environmental performance of its operations and products.**

inspiration> SC Johnson has committed to operating as responsibly as possible—reducing air emissions, water effluents, and solid waste as a ratio to its products, while decreasing use of fossil fuel energies and limiting greenhouse gas emissions.

achievements> Through its innovative Greenlist process, the company examines the environmental impact of each raw material used in its products, rating them on a scale of 3 to 0. An ingredient with a 3 rating is considered "Best," 2 is "Better," 1 is "Acceptable," while 0-rated materials are only used on a limited, approved basis.

SC Johnson has also phased out 100 percent of chlorine-based external packaging in its factories. In its top factories, it has employed innovative approaches to reduce greenhouse gas emissions by 5 percent annually since 2000, and achieved a 10 percent decrease in the use of fossil fuel energies—such as oil, coal, or natural gas.

staff> From the Chairman and CEO's office to the manufacturing line, SC Johnson recognizes that employees at all levels can contribute to and have an impact on the sustainability of its business, communities, products, and the planet.

suppliers> SC Johnson's Greenlist process was designed to allow flexibility that contributes to improved choices not only inside the company but also in its supply chain, rewarding the value of suppliers that demonstrate a higher level of environmental responsibility.

government> Recognizing the power of partnerships, in 2005, SC Johnson became the first major consumer packaged goods company to partner with the U.S. Environmental Protection Agency's Design for the Environment program. The company has also shared its Greenlist process with Environment Canada, the Chinese EPA, industry associations, universities, corporations, and others.

initiatives> The company is lending a hand to other companies on their path toward sustainability, with plans to license at no cost the Greenlist patent to companies committed to setting, tracking, and recording goals to reduce their environmental footprint.

challenges> While ahead of schedule on improving its overall Greenlist product score, there's more room for improvement as the company discovers new, more environmentally friendly raw materials for its products.

what's ahead> SC Johnson plans to continue to share expertise and serve as an example to other organizations through its website www.dowhatsright.com—a resource to educate companies about adding value and serving the greater good simultaneously.

Sourced from www.scjohnson.com

culture

the new CEO

31

Green business means putting environmental performance at the center of your working culture, expressed through your company values, business objectives, performance measures, and management structure. If not made a core part of your company's strategy, environmental goals can end up being tacked on to decisions after the financial and technical experts have had their day. Leading corporations are addressing this disconnect by appointing a chief environmental officer at the executive level, with authority equal to the chief financial officer, chief operations officer, or chief information officer. Any size business can profit from hiring an eco-leader. The point is not to create a green bureaucracy but to sharpen focus on implementation and show staff that sustainability is pivotal to business strategy.

Photo: Corbis Australia

green rewards

While consumers and investors increasingly demand better environmental and social performance from companies, the most important people in creating a green business culture are employees. Yet compensation and bonus plans still tend to be based on short-term measures such as quarterly share price and profit targets. Most management experts—not to mention directors and executives—agree these traditional financial measures shed little light on what creates long-term prosperity, and they can prevent people from seeing the benefit of pursuing sustainable business practices. From the front counter to the boardroom, create a culture that links salary, bonuses, and other rewards to achieving long-term environmental, social, and governance goals.

better suited

Business attire, particularly the traditional businessman's uniform of long-sleeved shirt, tie, and jacket, is ill-suited to the warmer regions, and to summer weather across much of the country. A formal business dress code for employees requires cranking up air conditioners to maintain comfortable conditions not only in your own workplace but in shops and food outlets as well. Start with casual Friday, and then adopt a policy that permits more climate-appropriate clothing—short-sleeved open-necked shirts, for instance—so your office thermostat can be set to a slightly higher temperature. Every degree you save will cut up to 20 percent from air conditioning costs. Provide closet space where staff can keep more formal attire for the occasions when it's necessary.

Photo: APL

uniform action

If your business uses uniforms, choose a material and supplier that ensure both physical and moral comfort. Synthetic textiles made from fossil fuels are non-biodegradable. Nylon creates nitrous oxide, a greenhouse gas 300 times more potent than carbon dioxide, and polyester requires large amounts of water. Rayon is made from wood pulp treated with hazardous chemicals. Natural fibers are also problematic. Cotton is the world's most chemical-intensive crop, requiring 10 to 18 applications of herbicides, insecticides, and fungicides along with nearly 3,500 gallons of water per pound. Wool requires over 20,000 gallons of water per pound, with sheep destroying natural habitat and emitting methane gas. Chemical-free, organic cotton, wool, and hemp are the best options.

flextime

35

The Texas Transportation Institute estimates that U.S. drivers waste nearly 6 billion gallons of fuel each year due to traffic congestion, typically during peak morning and evening commute times as people drive to and from work. This not only costs commuters nearly eight working days in time each year, it also adds nearly 60 million tons of greenhouse gases annually to the atmosphere—along with noxious pollutants such as carbon monoxide, volatile organic compounds, and particulates. Flextime arrangements can help staff avoid traffic jams, reducing their emissions as well as saving time. Flextime also benefits those who walk, ride, or use public transportation. Your business will reap the benefit of less stressed workers grateful for the time not stuck in traffic.

telecommute

36

Working from home eliminates both the costs of commuter travel along with workplace energy use, which is often many times more than needed at home. Governments are increasingly promoting telecommuting to combat air pollution, with research indicating just a 4 percent change in traffic volume can be the difference between free-flowing traffic and gridlock. Like flextime, telecommuting isn't for every workplace or every person, but you can probably use it more than you do. Real estate savings alone often outweigh the costs. The challenge is to move away from a culture of attendance to one based on performance, assessing staff on the work they do rather than the time they're seen to be spending on it.

37

super choice

Socially responsible or ethical investment is a powerful driver of more sustainable business practices, channeling money away from problem sectors into productive, eco-friendly enterprise. Along with harnessing the power of your own business investments, make it easy for staff to choose retirement investments that provide for their future without harming future generations. The financial establishment now accepts socially responsible investment as smart investment, with ethical funds pointing to returns that match or outperform general funds. A range of products, encompassing different investment strategies, is available through industry mutual funds and specialist fund managers. The Social Investment Forum provides more information at www. socialinvest.org.

salary packaging

Photo: courtesy of the Australian Greenhouse Office,
Department of the Environment and Heritage

Many companies offer salary packages that include a car—with the bigger the salary package, the bigger the car. When running costs such as fuel, insurance, and parking are included, an employee is effectively being encouraged to drive rather than use more eco-friendly transportation. If you subsidize staff to drive, why should others doing the right thing by the environment be left to pay their own way? Make it a level paying field by providing equivalent financial incentives to walk, use public transportation, or ride a bicycle. Your business can save on parking space and get a fitter, healthier, more productive staff member.

39

promote
public transport

How often do you stand around trying to hail a cab to get across the city simply because it requires less thought than working out what bus, train, or subway to catch? When you factor in things like time spent parking, public transportation is often just as quick as driving, as well as being much cheaper. Encourage staff to use public transportation wherever possible, and take the hassle out of it by providing maps and timetables for local services in the staff room or on your intranet. Buy tickets in bulk for staff use rather than just shelling out for cabs. Encourage greater use of public transportation between home and work through interest-free loans for annual travel passes.

40
pool resources

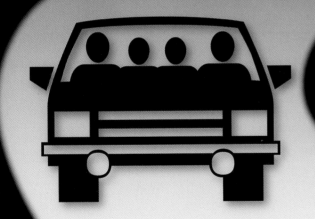

Three out of every four car trips transport just the driver. Encouraging carpooling is an obvious way to reduce carbon emissions created by commuting. Staff will save by sharing gas costs and your business will save on parking. Rather than just leaving it to informal arrangements, make carpooling part of your workplace culture: establish a database so people who work in different areas (or even neighboring businesses) can get in touch with one another. Provide incentives such as priority access to parking spaces, and a guarantee of transportation home so there are no worries about being stranded: experience shows the cost of doing so is negligible compared to the benefits.

< it is the creativity and enthusiasm of our staff that bring life to local initiatives >

MARRIOTT International runs more than 2,800 hotels in 68 countries, with 151,000 employees and revenue of about $10.4 billion. In 2006 it was named a U.S. Environmental Protection Agency Climate Leader for its pledge to reduce greenhouse gases by 6 percent per guest room by 2010. Mari Snyder is senior director of community relations.

inspiration> Many factors, beginning with founders J. Willard and Alice S. Marriott. Our efforts began in the early 1980s with energy-conservation programs driven primarily by rising energy costs.

achievements> The creation of our Environmentally Conscious Hospitality Operation in 1994 and Spirit To Serve Our Communities program in 1999. These programs formalize our commitment to sustainable business practices, communicate our signature issues and provide guidance to our local operations on how to get involved. We have an executive-level council to oversee environmental initiatives throughout the company.

management> These programs are aligned with our company's culture and make good business sense.

staff> We have received very positive responses to our programs from all our key stakeholders including our employees. They give their time and talent to help improve our environmental performance.

suppliers> We intend to take a more systematic review of our supply chain during the next stage of our journey toward sustainability.

government> In countries where governments are less supportive or lack acceptable guidelines we operate to our own international standards.

costs> Because energy and water conservation are among our major resource-use areas we are able to minimize expenses. Our social and environmental practices were a natural evolution in our company, which has always been focused on conserving resources and containing costs and has a strong culture rooted in hospitality and serving others.

benefits> While guidance and resources are provided on a global level, it is the creativity and enthusiasm of our staff that bring life to local social and environmental initiatives. In the past environmental initiatives and community engagement were encouraged but not mandatory. That is no longer the case. Environmental performance is becoming a business imperative.

initiatives> We have a range of energy- and water-saving programs. For instance our linen reuse program, which encourages guests to reuse linens and towels during their hotel stay, has saved water, detergents, and energy for heating. Replacing 4,500 outdoor signs with LED and fiber-optic technology cut their electricity consumption by 40 percent. New showerheads have reduced guests' hot-water consumption by 10 percent.

challenges> Change is always gradual. We have to first create awareness, implement, and then measure our progress. There may be issues that arise during the process but through benchmarking we can quickly identify and correct them through sharing best practices.

advice> 1> Adapt your business to take into account your stakeholders' interest in environmental and social issues. 2> Adopt a philosophy of "success is never final." 3> Focus on areas where you can make the greatest impact.

stakeholders

suppliers

Your business prosperity depends on many stakeholders—employees, customers, suppliers, contractors, regulators, and lobby groups. Think of it less as a chain than a web—one that sometimes demands a delicate balancing act. Consider your suppliers, for example. Just as you hope your own environmental efforts will help attract and retain customers, your own purchasing power is a means to influence the network of businesses you buy from. It may be necessary to swap to new suppliers that share and support your enthusiasm for green business practices, but that's not always possible or even desirable: there's also a place for persuasion and education, using your relationships to coax the less committed along the path to sustainability.

employees

As in nature, surviving and thriving in business requires the ability to learn and adapt. The importance of responding to feedback from customers doesn't need to be pointed out, but listening to other stakeholders is equally profitable. Employees can be sitting on a gold mine of suggestions for small, cheap, and easy-to-implement changes that can lead to big improvements in results. Set up a suggestion box, and acknowledge those who contribute, or adopt a more comprehensive strategy like quality circles— teams of workers (and others) who gather to discuss ways to improve processes. Pioneered in Japan in the 1960s, quality circles have been used to great effect by companies around the world to improve quality, lower costs, and motivate staff.

Photo: Corbis Australia

shareholders

43

enter

A nnual reports, chairman's statements, meeting notices, proxy forms, holding statements, and dividend advice—shareholder correspondence can add up to a mountain of paper, not to mention the resources and energy used to print and distribute it all. Replace printed with electronic communications whenever possible by sending documents and notices electronically. If you must print, double side pages and use software such as GreenPrint, which eliminates wasteful pages in any printout automatically, decreasing waste while letting you track the number of trees saved and your reductions in greenhouse gas.

44

customers

add to cart

From online newspapers to music and movies, more and more consumers are embracing the utility of the digital world. There are clear advantages to investing in efficient online transaction, delivery, and administration systems—not just in terms of convenience and speed of service for your customers but also in reducing the need for materials, transportation, and storefront space. To maximize the environmental pluses, however, you must also have good planning and management: internet shopping, for instance, won't necessarily contribute to lower transportation emissions if your deliveries are shipped from a far-flung distribution point.

community partnership

45

A ligning your business with community interests has measurable benefits, including enhanced reputation, greater customer loyalty, higher morale, and the ability to recruit and retain better people. Being community-minded, though, is no longer just about writing a check (as appreciated as that is). Visionary companies and community groups now want more meaningful partnerships based on mutual exchange: on the one hand, your business has valuable managerial expertise in honing strategic focus and making the most of resources; on the other, a community group knows all about being mission-focused and how to deal with different stakeholders. A partnership may include product donation, financial assistance, and sharing technical expertise, depending on the size, scope, and focus of your business.

Photo: Corbis Australia

local engagement

More than one in five Americans donate to green groups annually, and it isn't just the largest and most well-known organizations like Sierra Club and Greenpeace to which they lend a hand. Alongside the high-profile lobbying (and fund-raising) efforts of the national and international groups are a multitude of local groups engaged in practical efforts like planting trees, rehabilitating public open spaces, recycling garbage, turning used cooking oil into biodiesel, and promoting renewable energy. No matter what the size of your business, there's a group that can benefit from your support and in turn help raise your profile in the community in which you operate. For a list of national, state, and regional organizations that run local groups, go to the National Wildlife Federation's Conservation Directory at www.nwf.org/conservationdirectory.

workplace giving

A workplace charitable giving program is a simple way a business can support a community group by channeling the charitable impulses of its employees. Under the federal tax code, your office can deduct regular contributions from the pre-tax salaries of participating staff and forward them to any organization that has nonprofit status. This means they won't have to wait until the end of the financial year to claim a tax refund, while their nominated beneficiary gets a guaranteed income stream minus the usual fundraising and administration costs. The most effective programs reflect the concerns of those giving the money, so give your staff the right to decide to whom they can donate.

volunteering

While your business can benefit from encouraging staff to give money to a good cause, it can reap even more rewards by encouraging them to give time. Voluntary work is not only worthy in itself but can also help develop those skills recognized as value creators in modern business—initiative, creativity, teamwork, and the ability to extrapolate from one experience to another. Make a real commitment to a culture of volunteerism by setting aside time—on Friday afternoons, for example—when staff can work individually or collectively on social projects. To ensure it isn't just seen as slack time, make the work part of performance reviews, with agreed goals and evaluation indicators.

Photo: Corbis Australia

49

report publicly

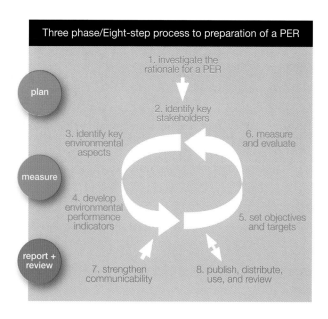

Three phase/Eight-step process to preparation of a PER

plan

measure

report +
review

1. investigate the
rationale for a PER

2. identify key
stakeholders

3. identify key
environmental
aspects

6. measure
and evaluate

4. develop
environmental
performance
indicators

5. set objectives
and targets

7. strengthen
communicability

8. publish, distribute,
use, and review

Publicly reporting your environmental performance is a crucial part of creating a sustainable business culture—and not just due to the expectations of customers, investors, and the wider community, but because of the internal focus it brings to operations. While protecting or enhancing business reputation is usually the initial motivation, many companies that publicly report say the real value has come from the discipline of developing benchmarks to measure their environmental performance, leading to better management of risks and opportunities. Employees are most frequently identified as the target audience. The Global Reporting Initiative provides an internationally used framework for public environmental reporting, and other useful resources, on its website: www.globalreporting.org.

join an
industry group 50

When the business community speaks, politicians listen. All too often, though, it is the noisy, self-interested few that get the attention of the media and thus of lawmakers. Being an active member of a green industry association can help address that imbalance. In the words of an old song, in unity is strength—the strength to push through the legal and regulatory changes needed at all levels of government for a future of sustainable economic development. There are also more direct benefits, such as networking opportunities, training programs, and marketing tools to assist you in growing your business.

< what is green? it's a question that has no single answer because every product has an environmental impact >

OFFICE DEPOT is a household word in offices around the globe, selling $15 billion of products and services to consumers and businesses of all sizes in the U.S. and internationally. Its massive operations consisting of office supply stores, a contract sales force, internet sites, direct marketing catalogs, and call centers—all supported by a network of crossdocks, warehouses, and delivery operations—provides more office products and services to more customers in more countries than any other company.

inspiration> Office Depot's environmental initiatives are driven by three environmental aspirations: to "buy green," to "be green," and to "sell green."

achievements> Office Depot's *Green Book*, available to contract customers and eligible organizations, makes it simple for customers to choose from thousands of green options among Office Depot's products. Office Depot now stocks more than 3,500 products with recycled content.

Office Depot's achievements extend beyond just what the company sells. The company is striving toward a goal of having one third of the paper it uses contain post-consumer recycled content. Office Depot has also spent millions in lighting retrofits, energy system upgrades, and air conditioner replacements, and it has entered into a contract to power several of its California stores with renewable energy, helping to offset its carbon emissions.

Office Depot has also had great success in selling remanufactured ink and toner cartridges, and in promoting recycling of empty cartridges, along with cell phones and rechargeable batteries.

suppliers> Office Depot is working with suppliers to increase the total volume of paper recovered from recycling, and the amount of this material—including material recovered from post-consumer waste fiber—in the paper products it distributes. The company pushes its suppliers to reduce pollution, including the phasing out of chlorine bleaching agents in the paper making process, and to source wood from forests independently certified to a third party standard.

partnerships> Office Depot's Conservation Alliance is the company's collaboration with three of the world's most respected, science-driven conservation organizations: Conservation International, NatureServe, and The Nature Conservancy. The company has turned to its alliance partners to help develop and implement an industry-leading environmental paper procurement policy.

what's ahead> Office Depot continues to encourage small steps toward greener offices around the world. To the company, that may mean even encouraging customers to take the first steps to convert from virgin paper (i.e., with no recycled content) to paper containing 10 percent post-consumer recycled content—starting them on the journey to a greener future.

> "Sustainability is a concept that touches each of our five corporate values: integrity, innovation, inclusion, customer focus and accountability."
>
> — Yalmaz Siddiqui, Office Depot Environmental Strategy Advisor

Sourced from www.officedepot.com

carbon neutral

calculate your footprint

51

By aiming to become carbon neutral you make your business stand out from the crowd, manage the risk of being left behind in an area set to become common practice, and position your business to profit from carbon trading. Before you can decide on the most effective way to become carbon neutral you need to measure accurately the greenhouse emissions caused directly by your own operations and indirectly by your suppliers and the use of your products. An assessment is best done with the help of an experienced carbon consultant and specialized assessor. For information on calculating your carbon footprint visit the website of the Greenhouse Gas Protocol (www.ghgprotocol.org), an initiative of the World Business Council for Sustainable Development.

green power

The U.S. has one of the highest per capita greenhouse gas emission rates in the world—and the problem is getting worse. By 2020 emissions from electricity could be as much as 160 percent higher than in 1990 (the benchmark year for the Kyoto Protocol). Yet for a carbon-conscious business the solution is simple. Your energy provider can source all or part of the electricity you use from renewable sources, such as solar, wind, and hydro. Though it costs a little more, your money supports crucial investment in renewable energy. Buying green power also entitles you to advertise your commitment and encourage other businesses to adopt more sustainable practices.

lighting

In many offices the lights burn long after employees have left the building. It might make for a pretty skyline but, combined with all those machines left on standby, it can double your energy bill. Lighting accounts for more than 20 percent of greenhouse gas emissions from commercial buildings. That figure could be reduced by as much as 70 percent through simple measures. By eliminating inefficient and unnecessary lighting you'll also reduce heat, reducing the cost of air conditioning. Ask your building manager to turn lights off at night or to install motion-activated sensors. Take the initiative by placing reminders near light switches in the area where you work. And if you're the last one out, flick the switch.

54

procurement

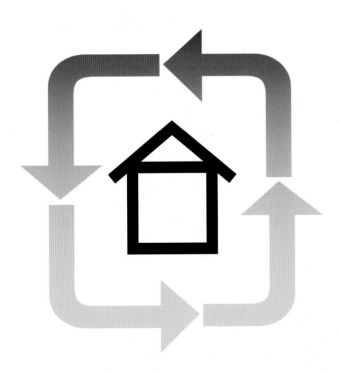

Importing goods from distant places where labor costs are cheaper and environmental and other regulation is less rigorous has its obvious attractions, but there's a price to pay once you start to factor in the cost of the carbon emissions generated through transportation. Make it part of your procurement policy to favor local trade. Apart from being good for the environment, there are many social benefits too. Local businesses are the backbone of the economy and of vibrant local communities. They create more local jobs, provide more scope for local producers, and result in more money being reinvested locally—including in your business.

vehicles

55

Greenhouse gas emissions from transportation—the second largest contribution to greenhouse emissions in the U.S.—are on the rise. The average fuel efficiency of vehicles in the U.S. is just over 25 miles per gallon, a figure that's changed little over the last 20 years. The private love affair with bigger, heavier cars and SUVs is only part of the problem: company fleet sales make up a significant portion of all new vehicles purchased. A website published by the U.S. Department of Energy and the Environmental Protection Agency, www.fueleconomy.gov, provides greenhouse gas and air-pollution ratings for all new vehicles. Visit greenfleets.org to get guidance on what you can do to cut fleet emissions and costs.

fleet fuel

A vehicle's eco-efficiency is improved by the fuel you put into it. High octane fuel, which contains up to a third less sulfur than regular unleaded gasoline, provides more engine power, more efficient consumption, and cleaner exhaust emissions. Look for gasoline blended with biofuels made from renewable or recycled sources, such as ethanol (made from corn or other biomass) and biodiesel (derived from vegetable oils or animal waste). Even a 10 percent ethanol blend will produce a third of the air pollutants of conventional gasoline. Buy established brands to be sure the fuel has been correctly blended at the refinery.

57

air travel

Jetting between cities for a meeting or conference is just about the single most damaging thing you can do for the environment. Air travel produces about as much carbon dioxide as each passenger driving his or her own car the same distance; and aircraft emissions, because they are released high in the atmosphere, have a greenhouse effect three times greater than road vehicle emissions. A single round-trip flight across the country, from New York to Los Angeles, contributes up to a ton in carbon dioxide emissions per passenger to global warming; the same distance in multiple short-haul flights produces even more due to the greater amount of fuel burned during take-offs and landings. Find ways to avoid any air travel that isn't absolutely necessary.

58

virtual meetings

In business "face time" is indispensable for your first meeting with an important client, trying to close a significant deal, asking an investor for money, or signing a contract. But for many other occasions meeting in person is a luxury rather than a necessity. Virtual meetings—teleconferencing, web-conferencing, or video-conferencing—can be just as effective, and cost a fraction of the time, money, and energy. The technology to enable virtual meetings is no longer exotic or expensive; in fact it can be free. If you have an internet connection the tools are just a few mouse clicks away.

59

carbon offset

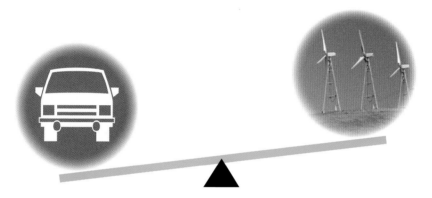

Carbon offsets can be used to complement direct reductions where it's impractical or uneconomic to further cut emissions from your business operations. For example, it may be impossible to reduce emissions further if you are already using the most energy-efficient equipment and vehicles available. The most positive strategy is to use a service like Carbonfund.org which offsets emissions through renewable energy, energy efficiency, and reforestation projects, providing much-needed investment in the technologies that can provide for a clean energy future. Consider tree-planting programs, which can absorb emissions as the trees grow, but only where there's a guarantee of the permanence of the trees and their benefit to local biodiversity.

be a climate leader

> > > > > >

The Climate Leaders program run by the U.S. Environmental Protection Agency provides guidance for businesses to accelerate energy efficiency, reduce greenhouse gas emissions, and promote awareness of greenhouse gas abatement opportunities. Nearly 140 business and industry groups, representing a broad cross-section of company type and sector, have joined the program. By participating in the program, your business gains access to advice from industry advisers as well as the opportunity to learn from the experience of other participants. Membership also entitles you to peer exchange opportunities and public recognition. For more information, go to www.epa.gov/climateleaders.

< you can make money without doing evil >

SUSTAINABILITY @ WORK

GOOGLE
Online search giant Google's mission is to organize the world's information and make it universally accessible and useful. The company's founders, Larry Page and Sergey Brin, developed a new approach to online searching from their Stanford University dorm room, quickly spreading their ideas to information seekers around the globe. Google is now widely recognized as the world's largest search engine—an easy-to-use free service that returns relevant results in a fraction of a second.

inspiration> Google's vibrant corporate culture—including games of beach volleyball, urban indoor workspaces infused with natural light, and an enthusiasm for the outdoors—is helping to transform how the company uses the sun in yet another way: by creating clean electricity.

achievements> Google has committed to solar energy production, launching the largest solar panel installation to date on a corporate campus in the U.S. Google has installed thousands of solar panels covering rooftops of eight buildings and two newly constructed solar carports at the "Googleplex" headquarters. The company also runs the largest corporate shuttle commuter program in the U.S., while offering employees cash incentives to purchase hybrid vehicles.

partnerships> Google.org, the philanthropic arm of Google, aspires to use the power of information to help people better their lives. In addition to financial resources, Google.org engages the company's entire family of people and partners, information technologies, and other resources to address three major growing global problems: climate change, global public health, and economic development and poverty.

initiatives> Google.org is hoping to give a boost to plug-in hybrid car technology through RechargeIT, an initiative that aims to reduce CO_2 emissions, cut oil use, and stabilize the electrical grid by accelerating the use of plug-in hybrid electric vehicles and vehicle-to-grid technology.

Along with the goal of being climate neutral by 2008, Google has also co-founded the Climate Savers Computing Initiative, a challenge to the industry to ramp up energy-efficient personal computers and server systems to 90 percent efficiency by 2010.

what's ahead> Google plans to invest approximately $10 million in technologies and companies featuring plug-in hybrids, fully electric vehicles, vehicle-to-grid capabilities, and batteries and other storage technologies. By investing in such green technology, Google hopes to help accelerate progress in addressing the climate and energy challenges of today's transportation sector.

> "One of the big objectives for Google as a corporation is to promote environmental protection."
>
> — Luanne Calvert, creative director, Google

Sourced from www.google.org

closing
the loop

call in the auditor

61

A waste audit helps your business understand the nature of its waste streams and the different options for transforming them into something useful. By closing the production loop you maximize resources, reduce waste-disposal costs, and create a more efficient organization. An audit involves collecting all waste generated over a week, then sorting through it to determine composition, quantities, and the effectiveness of existing waste management strategies. For best results you need well-defined objectives, committed management, and a plan that gives ownership to those who will implement systems and strategies. Contact your local solid waste utility or state environmental agency for advice and information about finding qualified waste auditors.

from output to input

Waste is simply that—a waste both of the world's resources and those of your business. Think about how you can create useable byproducts instead; transform something you would otherwise just throw away into something useful to your business or others. Used cooking oil from a fast-food restaurant, for instance, can be turned into biodiesel, a low-emission petroleum substitute that also takes care of the problem of how to dispose of the oil. Heat from machinery can be harnessed to meet hot water heating needs. Consider how you can alter your practices to substitute environmentally unfriendly inputs with recyclable and nontoxic ones.

63

collective action

One of the major obstacles to recycling is the cost-effectiveness of collecting it. By collaborating with other companies you can expand the potential for commercially attractive alternatives to landfills. Combine with neighboring businesses to collect quantities sufficient to interest a recycling service. Also seek out arrangements with other businesses that can use the products left over by your operations. Contact the U.S. Environmental Protection Agency for a list of international, national, and state-by-state materials exchange organizations at www.epa.gov/jtr/comm/exchange.htm.

Photo: Corbis Australia

plastics

64

The proportion of plastic recycled in the U.S. is only about 6 percent of the nearly 30 million tons thrown away each year. For some plastics—such as the widely used polyethylene terephthalate (PET)—recycling rates have actually declined in recent years. Local government recycling programs mostly collect PET and high-density polyethylene (HDPE); for other types of plastic you need to make your own recycling arrangements. If you're in the construction or food business, for instance, you probably use a lot of expanded polystyrene, which is completely recyclable but has a low recycling rate. For a list of plastic recyclers visit the Plastics Division of the American Chemistry Council's website on plastics and the environment at www.plasticsresource.com. Provide bins so staff can also recycle plastics that would otherwise end up in landfills.

phones

More than 140 million new cell phones are sold in the U.S. each year, yet their average life is less than two years. With handsets containing toxic metals—including arsenic, antimony, beryllium, cadmium, and lead—that adds up to a lot of potentially hazardous waste. Millions of old phones haunt home and office drawers because owners don't know what to do with them. Here's a solution—organize a workplace collection. Working phones can be sent to developing countries to help bridge the digital divide; otherwise they can be dismantled and their materials recovered to make other products. Phone recycler ReCellular collects and reprocesses more than 3 million retired phones annually.

Photo: Corbis Australia

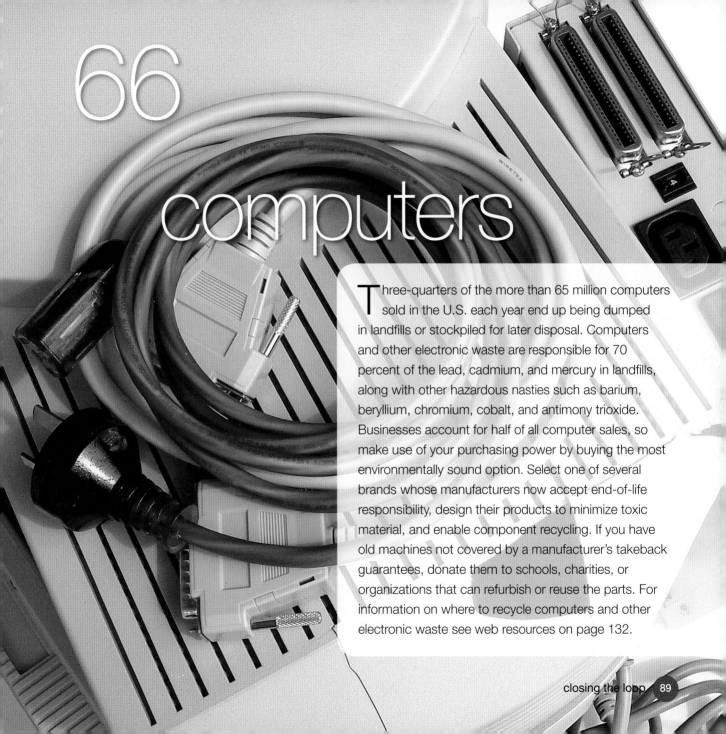

66

computers

Three-quarters of the more than 65 million computers sold in the U.S. each year end up being dumped in landfills or stockpiled for later disposal. Computers and other electronic waste are responsible for 70 percent of the lead, cadmium, and mercury in landfills, along with other hazardous nasties such as barium, beryllium, chromium, cobalt, and antimony trioxide. Businesses account for half of all computer sales, so make use of your purchasing power by buying the most environmentally sound option. Select one of several brands whose manufacturers now accept end-of-life responsibility, design their products to minimize toxic material, and enable component recycling. If you have old machines not covered by a manufacturer's takeback guarantees, donate them to schools, charities, or organizations that can refurbish or reuse the parts. For information on where to recycle computers and other electronic waste see web resources on page 132.

67

batteries

RECHARGE

Each year U.S. residents use nearly 3 billion dry-cell batteries to power everything from radios to phones, watches, laptops, and power tools. About 80 percent of batteries are single-use alkaline batteries, which while considered "safe" to dispose in landfills are hardly eco-friendly. Using rechargeable batteries can significantly reduce waste, plus save you money. But no battery lasts forever, and rechargeable ones contain more hazardous heavy metals and need to be recycled. The Rechargeable Battery Recycling Corporation provides guidance on the collection, storage, and recycling of batteries for both businesses and consumers (along with recycling schemes for phones) at www.rbrc.org.

food scraps

When food waste and other organic matter ends up in landfills and decomposes without air it can produce methane, a greenhouse gas 20 times more potent than carbon dioxide. It can also contaminate groundwater. Worm farms, available in a range of stylish designs suitable for offices, are a simple, effective, and clean way to convert food scraps into food for plants. Worms will munch through fruit, veggies, egg shells, and even pizza boxes (which can't be put in the paper-recycling bin) to produce mostly liquid waste and a smaller amount of solid waste that can be used to fertilize office plants. Each ton of organic matter you divert from the waste collection stream avoids over a third of a ton of greenhouse gases from landfills.

buy recycled

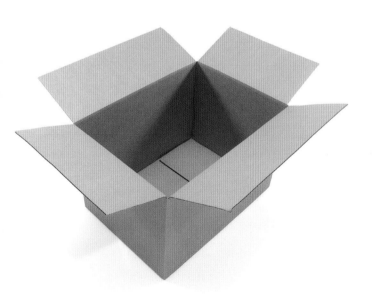

Collecting materials for recycling is one thing but your business isn't truly recycling unless it's buying recycled-content products. It's simple economics: along with the "push" from supply-side collection there must also be the "pull" from demand for recycled items. There are now ample choices in recycled content products available across several purchasing categories for your office, including binders, paper products, printer cartridges, and furniture. Information on these products is available through the U.S. Environmental Protection Agency's supplier database at www.epa.gov/epaoswer/non-hw/procure/database.htm.

Photo: Corbis Australia

life-cycle analysis
reuse repair
recycle

Budget-conscious businesses are always looking for the cheapest price, but the cheapest product is rarely the most cost-effective option. Such products are usually made with inferior parts that quickly wear out and cannot be replaced— a dubious design practice known as planned obsolescence that ensures continued demand at the cost of valuable resources and unnecessary waste. Cheap design also usually precludes effective recycling of components. Invest in making and buying well-designed, more durable items that can be repaired, upgraded, reused, and recycled. Ask your suppliers about their preparedness to take back products at the end of their life for reuse or recycling, along with what in-house practices they have adopted to improve their environmental performance.

< sustainability is key to staying competitive. sustainable practices drive efficiency >

VEOLIA Environmental Services (VES) provides waste-management services to 3,500 businesses and 550,000 households in Australia. The subsidiary of the French multinational Veolia Environnement has 2,300 staff and annual revenue of about $524 million. Tony Cade is group general manager for marketing and development.

inspiration> Improving the living environment is core to our business, therefore sustainable development is at the heart of everything we do.

achievements> Many, including partnering with the Australian Antarctic Division to remove and treat waste from contaminated sites in the sub-Antarctic and Antarctic regions, and investing in the Woodlawn Eco-Precinct, turning a disused open cut mine into a world-class resource and energy recovery center. In Australia we have already sold carbon credits generated from our operations to BP and HSBC in London.

management> VES is in the business of providing sustainable solutions. There was no resistance from the board.

staff> Our staff—and other stakeholders—are proud of our achievements. This is manifested not only in the company's strong financial performance but in the recognition of it as an employer of choice.

suppliers> We have streamlined our procurement practices with triple-bottom-line assessment criteria.

government> A more holistic "environmental scorecard" needs to be applied in developing legislative and regulatory measures. At the state level, for example, the use of a waste-diversion target (with accompanying levies) is too narrow and does not deliver a full suite of sustainable benefits. Cynics would suggest the levy is only there to generate income.

costs> Like any business, we need to meet financial hurdles when deciding on investments in alternative technologies/practices for waste-management. In years gone by there was resistance from some customers to use new "sustainable" methods in recovering resources based on a misconception that such approaches were more expensive. We are now seeing a shift in thinking such that all costs/risks are taken into account.

benefits> Our business is based on delivering triple-bottom-line benefits to customers and partners. Sustainability is key to remaining competitive. Apart from legislative imperatives, sustainable practices drive organizational efficiency.

initiatives> Too numerous to list, but one is the establishment of a specialized facility operations group that works with businesses to provide triple-bottom-line solutions that address operational, compliance, and environmental needs.

challenges> In the resource-recovery industry there have been some significant failures in alternative waste-treatment technology. The cost of failure can be high. The key lessons for any enterprise is to use proven technologies that have been validated at many other locations.

advice> 1> Adopt more holistic (triple-bottom-line) decision-making. Cheapest today is not always best or sustainable in the future. 2> Ensure that adopting sustainable practices is accompanied by a program for employee cultural change, developed in cooperation with service providers. 3> Have a comprehensive environment, health, and safety (EHS) program and ensure that service providers can meet (and exceed) enterprise expectations.

ecolabeling

ISO 14000

environmental assessment systems

ISO 14000 is the series of environmental assessment methods developed by the International Organization for Standardization, covering systems for environmental management, certification, and eco-labeling. The series defines three types of eco-labels. Type I labels (ISO 14024) are arguably the most valuable—for both producer and consumer—as they denote selective, multi-criteria-based, third-party-certified endorsement of a product. Type III labels (ISO 14025) provide quantified but non-selective product information based on independent verification against established benchmarks. Less valuable are Type II labels (ISO 14021), which are self-declared claims. Appreciating the relative merit of different labels will enable you to maximize the environmental benefit of your procurement decisions. For information about ISO 14000 go to www.iso.org. For more on eco-labeling go to www.gen.gr.jp, or eco-labels.org.

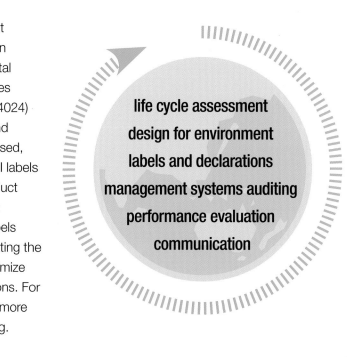

life cycle assessment
design for environment
labels and declarations
management systems auditing
performance evaluation
communication

green seal choice

There are several credible, independently certified eco-labeling programs in the U.S. covering consumer goods ranging from paper and wood to appliances and food. For a range of general office products, Green Seal's label certifies products that are environmentally preferable from a whole-of-life-cycle perspective compared to others in the same category. The Green Seal mark is awarded on independently audited and monitored criteria for environmental quality and social performance in conformity with ISO 14024 standards for Type 1 labels, covering products from paper to cleaners and machinery. Learn more at www.greenseal.org.

forest stewardship

As much as 30 percent of hardwood lumber imported into the U.S. comes from forests harvested illegally. Guarantee timber and wood products you buy come from legally harvested—and well managed—forests. The Forest Stewardship Council's FSC trademark identifies forest products containing verified FSC materials using a special label. Like Green Seal, FSC products are assessed and certified using social and environmental standards agreed to by the Forest Stewardship Council, an international coalition of timber buyers, traders, and non-government organizations—including environmental groups such as WWF and The Nature Conservancy. The FSC program has accredited more than 100 million acres of forests and several thousand products. For more information and a listing of FSC products visit www.fsc.org.

greenhouse friendly

Join the growing number of companies finding ways to make their products greenhouse-gas neutral from cradle to grave. Under an independent verification process, you can offset emissions from your production, use, and disposal practices through approved abatement projects. With the help of third party nonprofit organizations such as The Climate Trust and MyClimate as well as for profit companies, projects are generating permanent and verifiable greenhouse gas reductions—and helping to put the brakes on global warming. Projects include energy-efficiency upgrades, waste diversion and recycling, capture and flaring of landfill gas and other fugitive emissions, renewable energy generation, and tree planting projects.

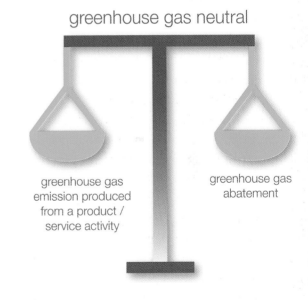

greenhouse gas neutral

greenhouse gas emission produced from a product / service activity

greenhouse gas abatement

energy star

Energy Star is an international standard for energy-efficient electronic equipment established by the U.S. Environmental Protection Agency. Energy Star-compliant machines reduce their power consumption by going to "sleep" when not being used and/or using less energy when in "standby" mode. This can save a lot of electricity on equipment like printers and fax machines, often idle about 95 percent of the time. Energy-saving features on machines like computers are not automatic so need to be activated. For information on the energy consumption of appliances while in low-power modes (including "off") go to www.energystar.gov.

water rating

It's now mandatory for all showerheads, faucets, toilets, and urinals sold in the U.S. to achieve a maximum flow rate in terms of water used per flush or gallons per minute. But for maximum water efficiency, select plumbing products with water-saving flow rates that exceed federal standards. Look for showerheads with flow rates of less than 2.5 gallons per minute, faucet aerators with flow rates below 1.0 gallon per minute, and toilets that don't exceed 1.6 gallons per flush. The U.S. Environmental Protection Agency's WaterSense program can help you find products independently verified and labeled as among the most water-efficient products in the marketplace.

fair trade

Guarantees a **better deal** for Third World Producers

FAIRTRADE ®

Though not specifically an eco-labeling system, Fair Trade certification promotes more sustainable agriculture by addressing exploitative trading practices that lead to abandonment of traditional farming practices, clearing of rain forests, and planting of single-species cash crops dependent on artificial fertilizers and pesticides. Fair Trade distributors (and ultimately consumers) pay a higher-than-market price for produce to ensure minimum labor, environmental, and social conditions are met. Fair Trade-labeled products are sourced directly from local cooperatives, putting more money in the pockets of growers. The International Fair Trade certification mark was launched in 2002 by Fairtrade Labeling Organizations International. For information on the labeling and certification system and on where to buy Fair Trade products go to www.fairtrade.com.

green globe

Green Globe is the worldwide benchmarking and certification program for the travel and tourism industry, including hotels, restaurants, resorts, and vehicle rental companies. The program measures performance in nine areas: greenhouse gas emissions; energy efficiency; freshwater use; waste water management; air quality protection and noise control; solid waste minimization, reuse, and recycling; ecosystem impact; land use; and local social, cultural, and economic impact. The certification process, developed in conjunction with Australia's Cooperative Research Centre for Sustainable Tourism, requires an on-site audit by an accredited third-party assessor. For more information and a list of participating businesses go to www.greenglobe.org.

sustainable responsible investment

Consider the many socially responsible investment options now available in the marketplace when developing your company's 401k program or other investments. When choosing ethically and socially responsible investment products and services, it pays to do your homework. You can research investment opportunities and find investment professionals who specialize in ethical investing through financial professional directories such as Socialfunds.com, Progressive Asset Management, or First Affirmative Financial Network. In doing so, you'll earn returns while using the power of your wallet to support a more sustainable future.

certified organic

Choose organic produce, meats, and other products that are grown or raised without synthetic chemicals or genetically engineered products, antibiotics, sewage sludge, irradiation, or growth hormones. Organic fruits and vegetables are grown with natural fertilizers and without pesticides, while animals are fed organic feed and allowed access to the outdoors. The environmental dividend is greater biodiversity at all levels of the food chain. The U.S. Department of Agriculture's national organic program includes an independently verified and fixed set of standards that must be met by anyone using the organic label in the U.S. You'll find the now familiar label on an ever-growing number of products in supermarkets, making your decision to go organic that much easier.

Photo: Corbis Australia

< …at the end of the day, our hope is that, through the people and places we touch, we do our share to create a better world and a more beautiful planet >

TIMBERLAND **has been making its trademark boots since 1973 and has grown from its origins as a small New England footwear company. Since then it's expanded to men's and women's clothing and other outwear and accessories. As the company has evolved, so too has its engagement with community causes, starting with a donation of 50 pairs of boots to the nonprofit group City Year in 1989. Since then, Timberland has emerged as a corporate leader in community service and environmental responsibility, investing employee time and money in communities and causes around the world.**

inspiration> Simply stated, Timberland's mission is to equip people to make their difference in the world.

achievements> Timberland uses an innovative "nutritional label" on its shoe and boot boxes, reporting on the company's social and environmental impact. Resembling government-mandated ingredient labels on food, box labels tell where the shoes or boots were made, how much energy was consumed to produce them, and how much renewable energy the company uses. The label also reports on the company's volunteer programs.

Inside each box, the company calls on its customers to take actions to help protect the environment or volunteer in their community. Footwear packaging uses soy-based inks and is made from 100 percent post-consumer recycled content cardboard. Timberland is also pushing to reduce its greenhouse gas emissions, increase the use of water-based adhesives in its shoes, and use more organic cotton.

staff> The company ensures that every new hire is passionate about corporate social responsibility, and all new corporate headquarters employees perform a day of volunteer service.

Serv-a-palooza, Timberland's daylong celebration of do-goodism, logs thousands of employee volunteer hours to combat social ills, help the environment, and improve conditions for laborers around the globe. The company's hybrid incentive provides $3,000 to any employee who purchases a hybrid vehicle.

suppliers> Timberland rates the environmental performance of each of the tanneries from which it buys leather, leading to substantial increases in tannery environmental performance since its launch. The company also monitors its suppliers to ensure they treat their workers fairly, and it monitors global human rights standards so its products are made in workplaces that are fair, safe, and nondiscriminatory.

government> Timberland has joined the Facility Reporting Project (FRP), an initiative of the socially and ethically responsible CERES coalition aimed at improving sustainability reporting and environmental performance at facilities across the country.

challenges> The company has invested $3.5 million in a solar array at its Ontario, California, distribution center. Although it will provide 60 percent of the center's energy, it may take as many as 20 years to show a return. But, according to CEO Jeffrey Swartz, the company is committed to sustainability over the long haul.

Sourced from www.timberland.com

marketing

spread the message

81

Surveys show more U.S. citizens are joining the growing ranks of environmentally concerned consumers around the world, with high expectations that companies be good corporate citizens. As the green market grows exponentially, leveraging your environmental and social commitments will help you gain and keep market share. Being "on-message" in your marketing is crucial. Start by sharing information with customers. Comprehensive product disclosure helps lead them through the maze of competing product claims and educates them about the impacts of their wider consumption. Empower them to make more informed purchasing decisions—not only when it comes to your product category but in others as well.

Photo: Corbis Australia

82

teaching aides

The size of the green market remains far smaller than what consumers' stated environmental concerns indicate it should be. This is a problem—and also an opportunity. It suggests many people are unsure about how to translate their personal values to their consumption choices, and perhaps also less than convinced about the difference it makes. Your staff plays a crucial role. They are at the front line in communicating the environmental performance of your business; they will reinforce the rest of your good work if they're able to "talk green" with demanding customers. Invest in education and training and the benefits will flow.

83

say something positive

If this were a tree, you wouldn't even notice.
Protect nature.

Marketing doesn't have to be just about encouraging people to buy stuff. The same principles used to sell products and services can also be used to sell ideas, attitudes, and behaviors. Marketing gurus Philip Kotler and Gerald Zaltman coined the term "social marketing" to describe such campaigns, in which the U.S. is among the world's leaders. They differ from other areas of marketing only in their objective: "Social marketing seeks to influence social behaviors not to benefit the marketer, but to benefit the target audience and the general society." But there's no reason you can't do both (as our pictured example shows). Integrate in your product pitch a useful social message.

stand out from the crowd

Selling behavioral change along with your own credentials becomes easier the more you integrate sustainability principles into all aspects of your business. Give green corporate gifts (such as organic wine and chocolates), send e-cards, and make your events carbon-neutral. Go even further by integrating environmental values into all your products. ShoreBank Pacific, the first commercial bank in the U.S. with a commitment to environmentally sustainable community development, offers not only individual savings and investment deposits, but lending programs that support individual and community efforts uniting conservation and economic development. With each loan, the bank also provides information on conservation improvements that can increase the value of the borrower's business.

Photo: Corbis Australia

85

show cause and effect

Donating a portion of your profits to a good cause is one of the most cost-effective ways to associate your business objectives with an environmental goal. Those simple words "a portion of proceeds go to …" can be a powerful influence when the usual core purchasing criteria—price, quality, appearance, taste, availability, and convenience—are less compelling. However, there can be criticism that you benefit more from the association than the cause you contribute to. Avoid such cynicism by backing up your pitch through the substance of your business practices, and be candid about the specific percentage of profits you are donating.

the whole package

86

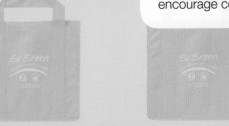

Packaging plays an important role in making products desirable. What it's made from and where it ends up also speaks volumes about your business's environmental commitment. More than 76 million tons of packaging and containers—over 500 pounds for every man, woman, and child—are thrown away in the U.S. each year; less than half of them are recycled. The rest ends up in landfills or, worse, as litter to be picked up by someone else. Reduce unnecessary packaging and take responsibility for its disposal. Seek biodegradable alternatives like bioplastic. Make it reusable and recyclable—and then include a message to encourage consumers to do so.

87

keep it real

From soft drinks to software, billions of dollars are spent on creating brand recognition through illusory feel-good associations. These campaigns can be fun but also seriously ironic. Consider the disconnect in alluring images of pristine rain forests being used to sell more SUVs, or postcard-perfect tropical islands to sell airline travel, or polar bears and other endangered wildlife to sell electronic goods. Nobody expects literal truth in all advertising, but obviously mixed messages can invoke a biting response from media-savvy consumers. Be a brand of the future by investing in honest associations.

Photo: Corbis Australia

beat the buzz

Research consistently shows that recommendations from peers have greater influence over purchasing decisions than all other forms of marketing—and the internet has extended the effect, giving individuals the power to share their views about a business not just with family and friends but an audience of hundreds, thousands, or millions. This is reflected in the statistics: those aged 18-34 rely more heavily on word of mouth than those aged 55 and over. One response is to increase your budget for strategies like affiliate and viral marketing, but the real solution is to focus on the underlying strength of any brand—authenticity. Quality, service, and reputation are what create sustainable sales. Let your company sell itself.

ink spots

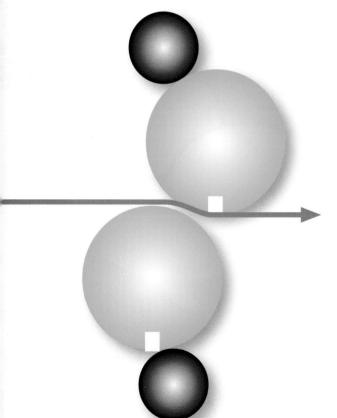

Despite the surging popularity of the internet, printed material remains a favored advertising form for many businesses. Yet Americans throw away unopened more than 44 percent of the direct mail they receive, and nearly 6 million tons of catalogs and other junk mail end up in landfills each year. If your marketing must rely on printed material the solution is simple: use an eco-conscious printer and demand eco-friendly inks and papers. For more information on green printing issues consult the Printers' National Environmental Assistance Center at www.pneac.org.

sharper vision

The link between printed material and the cutting of trees is reasonably obvious; but electronic forms of media can weigh as heavily. Radio, television, and even the internet contribute as much to climate change through their heavy energy use. Research has shown, for example, that Hollywood's contribution to air pollution in the Los Angeles region is second only to the petroleum industry. Seek to ensure the medium doesn't undermine your message. The commercial media depends on—and often only exists to attract—your advertising dollars. Use your influence to promote more sustainable practices. Have an environmental media policy that lays down minimum standards and encourages better performance.

< we believe in using business to inspire solutions to the environmental crisis >

PATAGONIA grew out of a small company making tools for alpine climbers. Founded by veteran climber Yvon Chouinard, alpinism remains at the heart of a company that's built a global name making clothes for outdoor adventure. Yet Patagonia's values and business practices still reflect those of a business started by a band of climbers and surfers with a deep reverence for outdoor exploration, a drive to do what's right, and a mission to protect the world's wild places.

inspiration> Patagonia's bold mission statement is to build the best product, do no unnecessary harm, and use business to inspire and implement solutions to the environmental crisis.

achievements> At every turn, Patagonia strives to minimize environmental impact in its buildings, using recycled materials, reclaimed wood, water-saving plumbing, and motion sensors to shut off unneeded lights. At headquarters, the company's own solar power plant and wind power help supply the company's energy needs. Patagonia also pledges 1 percent of its annual sales or 10 percent of pre-tax profits—whichever is more—to grassroots environmental organizations. The company even takes back its worn out clothing from customers for recycling.

staff> Few Patagonians are in it just for the money, though employees get an annual bonus based on profits. Employees can take off up to two months at full pay to work for environmental groups.

suppliers> Patagonia was the first major retail company to switch all of its cotton clothing to pesticide-free organic cotton, and the first to make fleece from recycled soda bottles.

government> Patagonia isn't waiting around for government to take the lead in conservation. In 1989, the company co-founded The Conservation Alliance to encourage other companies in the outdoor industry to give money to environmental organizations and to become more involved in environmental work.

costs> Long seen as a business maverick, Chouinard discovered that every time he made a decision because it was the right thing to do, Patagonia ended up making more money.

benefits> Chouinard says staying true to core values during thirty-plus years in business has helped create a company he's proud to run and work for. And the company's focus on making the best products possible has brought Patagonia success in the marketplace.

initiatives> Patagonia tackles various environmental campaigns each year, such as protecting the 1.5 million acre coastal plain of the Arctic National Wildlife Refuge by designating it wilderness.

challenges> Chouinard recognizes that everything the company does pollutes. Patagonia works tirelessly to reduce pollution in its products, processes, and facilities. Among its future challenges: to find technologies to make clothing out of synthetics that can be completely—and indefinitely—recycled into new products.

what's ahead> Patagonia's long-term goal? Take environmental responsibility for everything it makes.

Sourced from www.patagonia.com

green business

natural advantage

91

Business has always profited from nature, creating the conditions for more a comfortable existence by exploiting the earth's resources. When those resources seemed limitless the process of "taming" nature was acceptable, even necessary, but the conquest is turning into a hollow victory. Now, as Tachi Kiuchi and Bill Shireman write in *What We Learned in the Rainforest*, the real value of nature must come not from extracting its physical resources but from the lessons it teaches—and that includes how learning works. Take a cue from biologist Allan Wilson's studies of birds: create a learning organization by working together as individuals, not apart in isolated territories. Have a culture that encourages sharing information and new skills both within your own team and across the business.

Photo: Mallan Kyle

92

organic growth

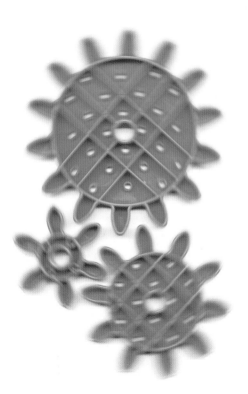

We like to think our organizations can run just like well-oiled machines, but that's aiming too low. Machine metaphors linger in business, from the clockwork universe to "cookie cutter" management. But though machine metaphors were undoubtedly useful in the past, they're wearing out. We are coming to appreciate that a business, and the economy as a whole, has more in common with a biological organism—a dynamic, interdependent system of living things that interact to produce a functioning, stable whole. As Arie de Geus, the former head of global planning for Royal Dutch/Shell, has put it: "Organizations are not rational." Gain greater efficiency from your business by shaping policies, structures, and systems to reflect that it is made up of people, not machines.

symbiosis

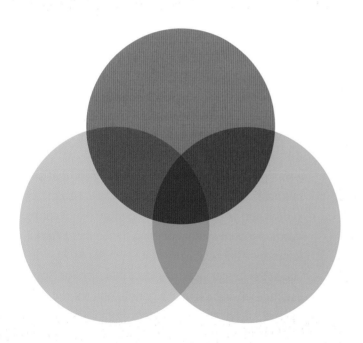

To serve the whole, a living community must serve its parts. Your business depends on the mutually beneficial interaction of its constituent elements—owners, employees, customers, suppliers, contractors, and the wider community. The idea it exists only to serve the owners' interests (as defined by corporate law, which places fiduciary duties to shareholders above all else) is ultimately destructive. The legal privilege of limited liability was originally extended only to enterprises deemed in the public interest. That idea is coming back. Green business regards profit as a means to an end, not the only end. Show the whole contribution of your business with accounts that measure the "triple bottom line" of financial, social, and environmental performance.

94

risk-aversion

Nature is defined by symbiosis, close associations of species that co-evolve into relatively stable and sustainable relationships. Even parasites usually avoid killing their hosts, since that would jeopardize their own existence. They meet the needs of the present without compromising the ability to meet needs in the future. So too in business: companies that survive over the long term are risk averse. They know that doing the right thing today is a matter of self interest. Not thinking of future consequences is a sure-fire recipe for eventual disaster, as lawsuits against companies for practices dating back decades now show. Just think of the movie *Erin Brockovich*. Plan for the long-term, not just the next year or quarter. It will lead to better, more strategic business—and lower insurance premiums.

		2007	2008	2009	2010	2011
2012	2013	2014	2015	2016	2017	2018
2019	2020	2021	2022	2023	2024	2025
2026	2027	2028	2029	2030	2031	2032
2033	2034	2035	2036	2037	2038	>>

95
mutualism

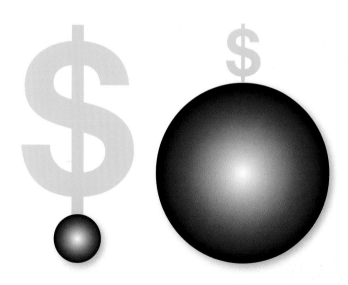

If the fair treatment of future generations—a principle known as intergenerational equity—is fundamental to environmental sustainability, then so too is the equitable treatment of the current generation. It is untenable for the richest 20 percent of the world's population to consume 80 percent of the world's resources. Without social sustainability there can be no environmental sustainability. Being a green business requires a commitment to both. Think of U.K.-based The Big Issue Foundation or Seattle's *Real Change* newspaper, providing jobs to the homeless, empowering them with income and employment opportunities. Or Grameen Bank's use of micro-credit, providing small loans to people so poor they are usually regarded as an unacceptable credit risk; loans of as little as $10 have empowered recipients to kick-start their own small businesses. Reflect on how your business can help bridge inequality.

96

biomimicry

Nature is a massive repository of design genius—from the spider that produces a material stronger and more flexible than Kevlar at room temperature to the bee whose fuel efficiency is more than 7 million miles per gallon of honey, both while performing valuable services to the ecosystem. Imitation of nature's designs and processes is being increasingly used in science and industry to solve human engineering, technological, and medical problems—from making nontoxic super-adhesives to treating multiple sclerosis. Biomimicry is the key to sustainable solutions for human problems as non-renewable resources become scarcer and more expensive to extract. For more examples of using nature as a model go to www.biomimicry. net. Be inspired and help preserve the database on which future economic prosperity depends.

97

innovation

Creating wealth from scarcity is a hallmark of natural systems. Consider the abundance of tropical rain forests, home to nearly half of the world's plant and animal species despite comprising less than a tenth of the earth's land surface. That rich biodiversity has nothing to do with the richness of the soil—it's actually very poor, as farmers have discovered to their misfortune after clearing the forest to plant single-species crops. The plants and animals have evolved through looking upward rather than downward for their resources. The lesson for business is obvious. Learning and adaptation, rather than reliance on abundant resources, is the true source of wealth. Invest in research and development along with education and training. Mine the value of your most unlimited resource: human intelligence and innovation.

Photo: Tim Wallace

conservation

In nature nothing is wasted; or to put it another way, all waste is food. When a plant or animal is eaten or dies, it's converted into resources that feed other species. This process was once depicted as a food chain—a linear diagram still often applied to business processes. But science has long since superseded the idea of the food chain with the food web or network, acknowledging the much more complex interactions used to feed resources back into the system in the most efficient manner. Innovative business is following suit, maximizing value by finding ways to feed materials back into the production cycle. As Bill Coors, the man who led the charge to replace one-use steel beer cans with recyclable aluminum ones, put it: "All pollution and waste is lost profit." Replace your linear production processes with nature-inspired closed loops for a healthier bottom line.

all waste is lost profit

self-organization

99

Nature thrives on distributed power systems. Consider the beehive—often thought of as a model of efficient hierarchical enterprise, of drones, workers, and above all the queen. In fact, bees teach a different lesson in corporate governance: the hive is both hierarchical and democratic, with many decisions made by consensus. When preparing to swarm, for example, the colony sends out scouts, then takes an "electronic" vote about which one to follow. From politics to electricity grids to the World Wide Web, efficient human systems are also based on distributed power. Business examples such as Semco in Brazil and Springfield Remanufacturing in the U.S. prove the success of "open-book management" and sharing ownership and control. Though it can be hard work and comes with its own challenges, empowerment is the more productive route for a successful business.

cooperation

We usually think of business as being all about competition. Actually we often think of nature that way, too—a war of all against all. But nature is both competitive and cooperative. As Lynn Margulis and Dorion Sagan write in *Microcosmos*: "Life did not take over the globe by combat but by networking." Species prosper through specialization, by finding niches and being useful to others: they serve themselves by serving the whole. Green business takes the hint, balancing its competitive spirit with the power of cooperation. Especially internally, a cooperative approach results in more productive and healthier outcomes. Share leadership, resources, and information. Let people flower and your business will reap the fruits.

Photo: Corbis Australia

< designs using nature's principles work better, are more beautiful, and are better value >

INTERFACE is the world's largest manufacturer of commercial and modular carpet, and widely acknowledged as a leader in industrial ecology. It employs about 4,800 staff with turnover of $111 million. Its vision is to be the world's first truly sustainable company, aiming to be environmentally restorative by 2020. **Robert Coombes** is president of Asia-Pacific operations.

inspiration> The path was formed by our chairman and founder, Ray Anderson, who was initially inspired by the book *The Ecology of Commerce,* by Paul Hawken.

achievements> We have reduced the company's carbon footprint by a third, adopted biomimicry in the design of products and processes; developed robust product stewardship programs, and shared our progress with other companies, leading to a wider sphere of influence.

management> Resistance exists until people grasp the fundamental premise that a sustainable business can be more competitive and more profitable.

staff> Because we care about *how* we make money, people want to work for us. We have higher staff engagement and lower staff turnover.

suppliers> Our procurement strategy dictates that suppliers helping to reduce any embodied environmental harm will receive more support. We have moved purchasing share to suppliers enabling our sustainability journey.

costs> You cannot create this degree of change in mentality without investing significant time and energy. Interface learned by doing and the results achieved brought people along the journey. A broad-based education program laid the foundation but the results achieved created the conviction and the passion for further progress.

benefits> Lower costs, better products, higher quality, more engaged people and more customers, who are increasingly looking for suppliers that can help them to embody a more sustainable footprint. Products designed using nature's principles work better and are more beautiful, and fewer extraneous features offer better value. Businesses that remain in disharmony with the natural world will ultimately fail—nature will dictate that. But before that happens, customers will dictate it.

initiatives> Sustainability champions are in place throughout the company to keep the rest of us honest. We encourage people to take responsibility and develop their own projects. Continuous measurement of the footprint of company sites and divisions helped spur a positive competitive drive to do well.

challenges> The journey is a long one and it is difficult to maintain the pace. We still have pockets of the company where the philosophy burns less bright and the job to improve is never complete. Senior leaders have to share the commitment and ensure it remains top of mind all the time.

advice> 1> It starts at the top. Unless the leadership is committed nothing will fundamentally change. 2> Focus on small steps. There are few silver bullets. Each small step well executed takes you somewhere else. 3> Begin with waste management: eliminate all activity the customer would not readily pay for. This creates a more efficient company with the capacity to invest in other sustainability initiatives.

resources

websites

Footprint Calculators	Carbon Footprint	www.carbonfund.org
	Carbon Calculator	www.americanforests.org
	Ecological Footprint	www.myfootprint.org
	Energy Calculators	www.eere.energy.gov/consumer/calculators
Water	Water-Savings Tips	www.h2ouse.net
	WaterSense	www.epa.gov/watersense
	Water Use It Wisely	www.wateruseitwisely.com
Energy	Alliance to Save Energy	www.ase.org/consumers
	Energy Savers	www.energysavers.gov
	Energy Star	www.energystar.gov
	Green-e Renewable Energy Certification	www.gree-e.org
	The Power is in Your Hands	www.powerisinyourhands.org
	U.S. Dept. of Energy, Office of Energy Efficiency and Renewable Energy	www.eere.energy.gov
Home and Garden	Coalition Against the Misuse of Pesticides	www.beyondpesticides.org
	Eco-Friendly Paints	www.eartheasy.com
	EPA, Indigenous Plants Landscaping	www.epa.gov/greenacres
	Gardens Alive!	www.gardensalive.com
	Gardener's Supply	www.gardeners.com
	USDA, Home Conservation Advice	www.nrcs.usda.gov/feature/backyard
New Homes	Build-e, Eco-Friendly Houses	www.build-e.com
	Certified Forests Products Council	www.certifiedwood.org
	Environmentally Construction Outfitters	www.environproducts.com
	Environmental Home Center	www.environmentalhomecenter.com
	No. American Insulation Manuf.'s Assoc.	www.naima.org
	U.S. Green Building Council	www.usgbc.org
Directory Services	EnviroLink Network	www.envirolink.org
	Green Pages Co-op	www.greenpages.org
	National Environmental Directory	www.environmentaldirectory.net
	Green Living Source for the Consumer	www.thegreenguide.com
At Work	Computer Recycling	www.computerrecyclingdirectory.com
	Conservatree	www.conservatree.com
	Office Footprint Calculator	www.thegreenoffice.com
	Reduce.Org	www.reduce.org

	Reducing Office Waste	www.filebankinc.com/reports/reduction_tips.htm
	The Real Earth, Inc.	www.treeco.com
Environmental Labeling	Ecolabels	www.eco-labels.org
	Electronic Product Environmental and Assessment Tool	www.epeat.net
	Energy-Star® Rating System	www.energystar.gov
	Global Ecolabeling Network	www.gen.gr.jp
	Green Seal Certified Cleaning Products and Paper Products	www.greenseal.org
	NSF International (Certification System)	www.nsf.org
	Transfair USA	www.transfairusa.org
Local Recycling	Nationwide Local Recycling Programs	www.earth911.org
	NSF's Recycling Guide	www.nsf.org/consumer/recycling
Computer Recycling	Earth 911	www.earth911.org
	Electronics Industries Alliance	www.eiae.org
Phone Recycling	Collective Good International	www.collectivegood.com
	The Charitable Recycling Program	www.charitablerecycling.com
	ReCellular	www.recellular.com
	Wireless recycling	www.wirelessrecycling.com
	Wireless Foundation	www.wirelessfoundation.org
Investment	First Affirmative Financial Network	www.firstaffirmative.com
	Grameen	www.grameen-info.org
	Progressive Asset Management	www.progressive-asset.com
	Social Funds/SRI World Group, Inc.	www.socialfunds.com
	Social Investment Forum	www.socialinvest.org
Food	Eartheasy	www.eartheasy.com
	Equal Exchange	www.equalexchange.com
	Green Restaurant Association	www.dinegreen.com
	Seafood Choices Alliance	www.seafoodchoices.com
	TransFair USA	www.transfairusa.org
	Whole Foods Market	www.wholefoods.com
Shopping	Co-op America, National Green Pages	www.coopamerica.org/pubs/greenpages/
	Earth Animal	www.earthanimal.com

websites *(continued)*

	Ecomall—Environmental Shopping Center	www.ecomall.com
	Global Exchange/Fair Trade	www.globalexchange.org
	One Percent for the Planet	www.onepercentfortheplanet.org
	Professional Wet-cleaning Network	www.tpwn.net
	Responsible Shopper	www.responsibleshopper.org
	Reusable Shopping Bags	www.reuseablebags.com
Transport	Center for Climate Change & Environmental Forecasting	www.climate.dot.gov
	Car Information—Mileage, Hybrids	www.fueleconomy.gov
	Electric Vehicle Assoc. of America	www.evaa.org
	Environmental Guide to Cars and Trucks	www.greenercars.com
	Transportation Almanac—Energy, Pollution	www.bicycleuniverse.info
For the Kids	Bobbie Big Foot	www.kidsfootprint.org
	Cool Kids for A Cool Climate	www.coolkidsforacoolclimate.com
	Earth Kids 911	www.earthkids911.org
	Water Use Calculator	www.ga.water.usgs.gov/edu/sq3.html
Sustainable Lifestyles	American Forests	www.americanforests.org
	Conservatree	www.conservatree.org
	Earth Easy	www.eartheasy.com
	Earth 911	www.earth911.org
	Forest Stewardship Council	www.fsc.org
	Green Globe	www.greenglobe21.com
	National Arbor Day Foundation	www.arborday.org
	National Parks Conservation Association	www.npca.org
	Trees for the Future	www.treesftf.org
	U.S. Green Building Council	www.usbg.org
Advocacy Groups and Organizations	American Council for An Energy-Efficient Economy	www.aceee.org
	Blue Ocean Institute	www.blueocean.org
	Certified Humane	www.certifiedhumane.com
	Children's Health Environmental Coalition	www.checnet.org
	Earthshare	www.earthshare.org
	Environmental Defense Fund	www.environmentaldefense.org
	Environmental Protection Agency	www.epa.gov
	Environmental Working Group	www.ewg.org
	Friends of the Earth	www.foe.org
	Green Peace USA	www.greenpeace.org

	Idealist	www.idealist.org
	Marine Stewardship Council	www.msc.org
	Natural Resources Defense Council	www.nrdc.org
	Rainforest Action Network	www.ran.org
	Redefining Progress	www.rprogress.org
	Rocky Mountain Institute	www.rmi.org
	Social Marketing Institute	www.social-marketing.org/index.html
	Stop Global Warming	www.stopglobalwarming.org
	The Conservation Fund	www.conservationfund.org
	The Nature Conservancy	www.nature.org
	The Ocean Conservancy	www.oceanconservancy.org
	Worldwatch Institute	www.worldwatch.org
	World Resources Institute	www.wri.org
	World Wildlife Fund	www.wwf.org.
Media	E/The Environmental Magazine	www.emagazine.com
	Earth Policy Institute	www.earth-policy.org
	Environmental Issues Newsletter	www.environment.about.com
	Environmental Health News	www.environmentalhealthnews.org
	Environmental News Network	www.enn.com
	Grist Magazine	www.grist.org
	Tree Hugger, Online Magazine	www.treehugger.com
Government Agencies	EPA Climate Leaders	www.epa.gov/climateleaders
	Materials Exchange Resources	www.epa.gov/jtr/comm/exchange.htm
	U.S. Environmental Protection Agency, Energy Star Program	www.energystar.gov
	U.S. Environmental Protection Agency, Waste Wise Program	www.epa.gov/epaoswer/non-hw/reduce/wstewise
	USDA National Organic Program	www.ams.usda.gov/nop
	WaterSense	www.epa.gov/watersense
Greenhouse gases	Greenhouse Gas Protocol	www.ghgprotocol.org
	The Climate Trust	www.climatetrust.org
Procurement	Green Seal	www.greenseal.org
	Suppliers of Recycled Content Products	www.epa.gov/epaoswer/non-hw/procure/database.htm
Carbon offsetters	Carbonfund	www.carbonfund.org

glossary

biodegradable> capable of decaying as a result of the action of micro-organisms that break the material down into naturally recyclable elements.

biodiversity> all life on earth, including the variability within and between ecological communities or systems.

biofuel> low-emission petroleum substitute made from renewable or recycled sources; includes ethanol made from cereals and sugarcane and biodiesel made from vegetable oils or animal waste.

biomimicry> imitating biological solutions in the design of products and processes.

bioswale> landscape element designed to capture rainwater runoff and filter silt and pollution before it enters the water table.

carbon footprint> the amount of carbon dioxide emitted into the atmosphere by an individual, organization or nation; usually measured in metric tons of carbon dioxide emitted annually.

carbon neutral> achieving zero net greenhouse emissions through activities to reduce gas creation and paying for offsetting schemes such as renewable energy generation and tree-planting.

carbon sequestration> a process to reduce carbon in the atmosphere; a variety of sequestration methods, including removing carbon from flue gases then storing it underground, are being explored.

carbon sink> any mechanism that removes greenhouse gases or aerosols from the atmosphere. Oceans, trees and soil are all carbon sinks.

carbon trading> a market to encourage carbon abatement through selling and buying "carbon credits," enabling heavy emitters to meet reduction commitments by buying surplus emission reductions. Trading markets are well-established in Europe and the United States.

climate change> changes to planetary climatic conditions as a result of global warming; scientific consensus links warming to the extra carbon dioxide and other greenhouse gases added to the atmosphere by human activity.

corporate social responsibility> the concept that all companies, regardless of ownership structure, have obligations to customers, employees, and the wider community. The idea of good corporate citizenship extends further than the obligation to obey all laws.

cradle to cradle> an extension of the concept of cradle to grave aimed at designing and making products so efficiently that they can benefit the earth through their use and disposal.

cradle to grave> the whole of a product's life cycle, from the extraction and processing of natural resources used to manufacture it through to the resources consumed in transportation, use and ultimate disposal.

dioxin> the popular name for a family of organic compounds that bio-accumulate with toxic effect in humans and wildlife. Two of the most widely studied sources of dioxins are the making of the herbicide Agent Orange and the chlorine bleaching of wood pulp in paper-making.

ecological footprint> the estimated amount of productive land required to provide all the resources consumed and absorb all the wastes created by an individual or group. The calculation is usually expressed in terms of how many planets would be needed for all the world's population to live the same way.

environmental audit> an assessment of the environmental impacts of an organization's operations.

ethical investment> a range of investment strategies that incorporate environmental and social measures alongside financial returns; also known as socially responsible investment (SRI).

e-waste> discarded electrical equipment such as mobile phones, computers, DVD players, and cabling.

extended producer responsibility> a policy approach in which a producer's responsibility for a product is extended to the post-consumer stage of the product's life cycle.

greenhouse abatement> activity that contributes to the reduction of greenhouse gas emissions.

greenhouse gas> any atmospheric gas that contributes to the greenhouse effect by absorbing energy from the sun. Naturally occurring gases include water vapor, carbon dioxide, methane, nitrous oxide and ozone. Human activities add to these gases.

green power> electricity generated from renewable sources such as hydro, wind and solar, avoiding the emissions associated with the burning of fossil fuels.

high-octane fuel> a gasoline formula that releases more energy when combusted. Its refinement significantly reduces the amount of sulfur compared to regular petrol, providing cleaner exhaust emissions along with more engine power and lower fuel consumption.

hybrid engine> an engine that combines both a petrol and an electric motor, capturing the energy used in braking and dramatically reducing fuel consumption.

intergenerational equity> the fair treatment of future generations.

Kyoto Protocol> international agreement on global warming and emissions targets set at the United Nations Conference on Climate Change in Kyoto, Japan, in 1997. The United States and Australia are the only two industrialized nations to not have ratified.

landfill> disposal of solid waste by burying it between layers of dirt in low-lying ground or excavated holes.

life-cycle assessment> the study of all inputs and outputs of materials and energy to determine the environmental impact attributable to the functioning of a product or service over its whole life time.

methane> a gas with a greenhouse effect 23 times greater than carbon dioxide. Methane is produced naturally, including from volcanoes, wetlands, termites and the ocean, and by human activity, including from flatulent livestock and the decomposition of organic matter buried in landfill.

organic> produced without the use of fossil-fuel based fertilizers, synthetic pesticides, or genetically modified crop varieties.

plastic code> A number identifying the most common plastic type in a product or packaging. All plastics marked 1 to 7 are theoretically recyclable though in practice many are not: 1> polyethylene terephalate (PET); 2> high density polyethylene (HDPE); 3> unplasticized polyvinyl chloride (UPVC) or plasticized polyvinyl chloride (PPVC); 4> low density polyethylene (LDPE): 5> polypropylene (PP); 6> polystyrene (PS) or expandable polystyrene (EPS); 7> other, including nylon and acrylic.

RCP> a recycled content product, containing pre-consumer recycled content from scraps, trimmings, and overruns left over from manufacturing processes and/or post-consumer recycled content from products that have served their original use.

sustainability> the ability to meet the needs of the present without compromising the ability to meet the needs of the future.

thermal mass> a structure capable of absorbing and storing heat energy, leading to increased comfort for building occupants and a reduced need for artificial heating and cooling.

volatile organic compound> a chemical that can vaporize and enter the atmosphere under normal conditions. Trees are the major outdoor source of VOCs. Paints, cleaning agents, and furnishings made from petrochemicals are major sources of indoor VOC pollution.

waste management> action to reduce waste going into landfill, through material efficiency, waste wreduction and the recovery and reuse of discarded materials.

Clean up
the world

about Clean Up the World

Clean Up the World, the international outreach campaign of Clean Up Australia, was co-founded by *True Green* creator Kim McKay and Ian Kiernan, AO—legendary yachtsman and 1994 Australian of the Year.

In partnership with the United Nations Environment Programme (UNEP), Clean Up the World annually attracts more than 35 million volunteers who join community-led initiatives to clean up, fix up, and conserve their local environment.

Fifteen years after its launch, the campaign has become a successful action program that spans more than 120 countries, encouraging communities to take control of their own destiny by improving the health of their community and environment.

Global activities include waste collection, education campaigns, environmental concerts, creative competitions, and exhibitions on improving water quality, planting trees, minimizing waste, reducing green house gas emissions, and establishing recycling centers.

Participants range from whole countries (e.g. Australia and Poland), community and environmental groups, schools, government departments, businesses, consumer and industry organizations, to sponsors and dedicated individuals who either work independently in their local communities or with other groups in a coordinated effort at a regional or national level.

Visit the Clean Up the World website to find out how your community, company, or organization can become involved: **www.cleanuptheworld.org**

"For more than 15 years Clean Up the World has empowered individuals to take care of our environment. The work of our volunteers has made and will continue to make significant inroads, but now it's time to move to the next stage and address the significant environmental threats that face us today in the key areas of climate change, waste and water."

Ian Kiernan, AO
Chairman & Founder, Clean Up the World

Photo: Marc Stanley, titomedia

Kim McKay (right) is the co-founder and deputy chairwoman of Clean Up Australia and Clean Up the World. She is an international social marketing consultant who counts National Geographic among her clients.

Jenny Bonnin (left) is a director of Clean Up Australia and Clean Up the World. She and Kim are partners in the social marketing firm, Momentum2. Jenny has two children and lives with her partner and extended family.

Their previous book, *True Green: 100 Everyday Ways You Can Contribute to a Healthier Planet* was released in Australia in 2006 and was published in the United States in April 2007.

writer
Tim Wallace is a freelance business journalist and editor of ecologicmedia.com. He has worked for the *Australian Financial Review*, *AFR Boss* magazine, *The Sydney Morning Herald,* and *The Age* and has written on business issues for *The Diplomat* magazine.

designer
Marian Kyte is a freelance designer with a passion for incorporating sustainability principles in her work. Her clients have included Qantas, Craftsman House Books, Power Publications, Sherman Galleries, *Art & Australia,* *Limelight* magazine, and *True Green*.

acknowledgments

True Green @ Work is a true team effort and we are indebted to Tim Wallace for his amazing knowledge, insights, and turn of phrase, and to Marian Kyte for her sheer brilliance as a designer. Special thanks to journalist Dave Wortman for revising the text and providing additional research in the U.S. We'd also like to thank Tania Baxter for her coordination and online work, and Kerry McKay and Jessamine Walker for their support.

We greatly admire and appreciate the support of Molly Harriss Olson from the Business Leaders Forum on Sustainable Development in Australia. She is unrelenting in her passion and commitment to protect the environment and has played an important role on a global scale influencing business to improve its behaviour. Ian Kiernan is a true friend who doesn't ever give up and is a great inspiration to us all.

We also wish to thank Nina Hoffman, Kevin Mulroy, Olivia Garnett, and Lauren Pruneski from National Geographic Books for their support.

We dedicate this book to our parents, Geoff and Sheila Crow and Gordon and Coral McKay, who set us on the right paths in life.